# HOW THE
# UNIVERSE
# GOT ITS SPOTS

# HOW THE
# UNIVERSE
# GOT ITS SPOTS

DIARY OF A FINITE TIME IN A FINITE SPACE

## Janna Levin

Weidenfeld & Nicolson

LONDON

First published in Great Britain in 2002 by
Weidenfeld & Nicolson

© 2002 by Janna Levin

A CIP catalogue reference for this book
is available from the British Library

ISBN 0 297 64651 6

Typeset by Selwood Systems, Midsomer Norton

Printed in Great Britain by
Butler & Tanner Ltd, Frome and London

Weidenfeld & Nicolson

The Orion Publishing Group Ltd
Orion House
5 Upper Saint Martin's Lane
London, WC2H 9EA

# CONTENTS

*Acknowledgements*    vii

*Preface*    ix

1 Is the universe infinite or is it just really big?    1

2 Infinity    5

3 Newton, 300 years and Einstein    16

4 Special relativity    23

5 General relativity    37

6 Quantum chance and choice    50

7 Death and black holes    63

8 Life and the big bang    79

9 Beyond Einstein    99

10 Adventures in Flatland and hyperspace    104

11 Topology: links, locks, loops    115

12 Through the looking glass    131

13 Wonderland in 3D    141

14 Mirrors in the sky    151

15 How the universe got its spots    162

16 The ultimate prediction    178

17 Scars of creation    185

18 The shape of things to come    194

*Epilogue*    199

*Index*    201

# ACKNOWLEDGEMENTS

I am so grateful to everyone who took care of me in California and in New York especially Angelina de Antonis, Jason Coleman, Alene Dawson, Nancy Eastep, Sean Hayes, Eno Jackson, Rory Kelly, Prudence Longaker, Sean McGuire, Sylvie Myerson, Diane Olivier, Ruthonly, Sara Jane Parsons, Karen Rait, Andy Rasmussen, Will Waghorn, the San Francisco drawing group, and everyone in the Oakland commune. Thanks to Warren Malone for providing so much material and to all the support in London from Bergit Arends, Jaki Arthur, Paul Bonaventura, Bernard Carr, Sarah Dunant, Siân Ede, Pedro Ferreira, Jem Finer, Na'ama Gidron, Jonathan Halliwell, Annabell Huxley, Chris Isham, Mark Lythgoe, Joao Magueijo, Sallie Robbins, Valerie Rosewell, Lee Smolin, Richard Wentworth, Tom Wharton, Pitt Wuehrl, PPARC, DAMTP, the CfPA, and the Theoretical Physics Group at Imperial College, the sci/art community, and everyone on the fourth, Brian Deegan, Eric Jorrin, Whitney Hanscom, Ben McLaughlin, Tim Williams and Blast Theory, and to my friends and colleagues I have worked with and who have taught me so much about topology and cosmology, John Barrow, Dick Bond, Neil Cornish, Giancarlo de Gasperis, Imogen Heard, Jean-Pierre Luminet, Dmitry Pogosyan, David Spergel, Glenn Starkman, Evan Scannapieco, Joe Silk, George Smoot, Tarun Souradeep and Jeff Weeks. Forgive me anyone I have carelessly omitted. I am especially grateful to my editor Peter Tallack for his insight and vision. I don't know how to acknowledge the seemingly endless support of my family. Thank you Leslie Levin for not letting me back down. Thank you John Hibbard, Ari and Jack Hibbard, Stacey and Cami Levin, the Jacobsons, the Kavins, the Levins and Eve Jacobson, and most of all Sandy and Richard Levin.

# PREFACE

These letters were originally written to Sandy Levin, my mom, my friend. They've mutated and grown into a two-year diary of unsent letters. Although I don't address her very directly, the second person singular 'you' when it appears is a reference to Sandy. I confess that this is not an historical treatise and I fail terribly at citing the many tremendous people who strove to build these great ideas.

For anyone wanting a less personal account of life in the cosmos, there are many excellent books. My very incomplete list of such good books includes: *The Artful Universe, Theories of Everything* and *Pi in the Sky* by John D. Barrow, *The Cosmic Code* by Heinz Pagels, *The First Three Minutes* by Steven Weinberg, *Gödel, Escher, Bach: The Eternal Golden Braid* by Douglas Hofstadter, *Lonely Hearts of the Cosmos* by Dennis Overbye, *The Elegant Universe* by Brian Greene, *Black Holes and Time Warps: Einstein's Outrageous Legacy* by Kip Thorne, *A Brief History of Time* by Stephen Hawking, *The Life of the Cosmos* and *Three Roads to Quantum Gravity* both by Lee Smolin. These books are excellent pedagogical references for anyone who wants to learn in more detail about some of the ideas that are only touched upon in this diary.

# I

## IS THE UNIVERSE INFINITE
## OR IS IT JUST REALLY BIG?

Some of the great mathematicians killed themselves. The lore is that their theories drove them mad, though I suspect they were just lonely, isolated by what they knew. Sometimes I feel the isolation. I'd like to describe what I can see from here, so you can look with me and ease the solitude, but I never feel like giving rousing speeches about billions of stars and the glory of the cosmos. When I can, I like to forget about maths and grants and science and journals and research and heroes.

Boltzmann is the one I remember most and his student Ehrenfest. Over a century ago the Viennese-born mathematician Ludwig Boltzmann (1844–1906) invented statistical mechanics, a powerful description of atomic behaviour based on probabilities. Opposition to his ideas was harsh and his moods were volatile. Despondent, fearing disintegration of his theories, he hanged himself in 1906. It wasn't his first suicide attempt, but it was his most successful. Paul Eherenfest (1880–1933) killed himself nearly thirty years later. I looked at their photos today and searched their eyes for depression and desperation. I didn't see them written there.

My curiosity about the madness of some mathematicians is morbid but harmless. I wonder if alienation and brushes with insanity are occupational hazards. The first mathematician we remember encouraged seclusion. The mysterious Greek visionary Pythagoras (about 569 BC–about 475 BC) led a secretive, devout society fixated on numbers and triangles. His social order prospered in Italy millennia before labour would divide philosophy from science, mathematics from music. The Pythagoreans believed in the mystical meaning of numbers and developed a religion tending towards occult numerology. Their faith in the sanctity of numbers was shaken by their own perplexing mathematical discoveries. A Pythagorean who broke his vow of secrecy and exposed the enigma of numbers that the group had uncovered was drowned for his sins. Pythagoras killed himself too. Persecution may have incited his suicide, from what little we know of a mostly lost history.

When I tell the stories of their suicide and mental illness, people always wonder if their fragility came from the nature of the knowledge – the knowledge of nature. I think rather that they went mad from rejection. Their mathematical obsessions were all-encompassing and yet ethereal. They needed their colleagues beyond needing their approval. To be spurned by their peers meant death of their ideas. They needed to encrypt the meaning in others' thoughts and be assured their ideas would be perpetuated.

I can only write about those we've recorded and celebrated, if posthumously. Some great geniuses will be forgotten because their work will be forgotten. A bunch of trees falling in a forest fearing they make no sound. Most of us feel the need to implant our ideas at the very least in others' memories so they don't expire when our own memories become inadequate. No one wants to be the tree falling in the forest. But we all risk the obscurity ushered by forgetfulness and indifference.

I admit I'm afraid sometimes that no one is listening. Many of our scientific publications, sometimes too formal or too obscure, are read by only a handful of people. I'm also guilty of a self-imposed separation. I know I've locked you out of my scientific life and it's where I spend most of my time. I know you don't want to be lectured with disciplined lessons on science. But I think you would want a sketch of the cosmos and our place in it. Do you want to know what I know? You're my last hope. I'm writing to you because I know you're curious but afraid to ask. Consider this a kind of diary from my social exile as a roaming scientist. An offering of little pieces of the little piece I have to offer.

I will make amends, start small, and answer a question you once asked me but I never answered. You asked me once: what's a universe? Or did you ask me: is a galaxy a universe? The great German philosopher and alleged obsessive Immanuel Kant (1724–1804) called them universes. All he could see of them were these smudges in the sky. I don't really know what he meant by calling them universes exactly, but it does conjure up an image of something vast and grand, and in spirit he was right. They are vast and grand, bright and brilliant, viciously crowded cities of stars. But universes they are not. They live in a universe, the same one as us. They go on galaxy after galaxy endlessly. Or do they? Is it endless? And here my troubles begin. This is my question. Is the universe infinite? And if the universe is finite, how can we make sense of a finite universe? When you asked me the question I thought I knew the answer: the universe is the whole thing. I'm only now beginning to realize the significance of the answer.


**3 SEPTEMBER 1998**

Warren keeps telling everyone we're going back to England, though, as you know, I never came from England. The decision is made. We're leaving California for England. Do I recount the move itself, the motivation, the decision? It doesn't matter why we moved, because the memory of why is paling with the wear. I do remember the yard sales on the steps of our place in San Francisco. All of my coveted stuff. My funny vinyl chairs and chrome tables, my wooden benches and chests of drawers. It's all gone. We sit out all day as the shade of the buildings is slowly invaded by the sun and we lean against the dirty steps with some reservation. Giant coffees come and go and we drink smoothies with bee pollen or super blue-green algae in homage to California as the neighbourhood parades past and my pile of stuff shifts and shrinks and slowly disappears. We roll up the cash with excitement, though it is never very much.

When it gets too cold or too dark we pack up and go back inside. I'm trying to finish a technical paper and sort through my ideas on infinity. For a long time I believed the universe was infinite. Which is to say, I just never questioned this assumption that the universe was infinite. But if I had given the question more attention, maybe I would have realized sooner. The universe is the three-dimensional space we live in and the time we watch pass on our clocks. It is our north and south, our east and west, our up and down. Our past and future. As far as the eye can see there appears to be no bound to our three spatial dimensions and we have no expectation for an end to time. The universe is inhabited by giant clusters of galaxies, each galaxy a conglomerate of a billion or a trillion stars. The Milky Way, our galaxy, has an unfathomably dense core of millions of stars with beautiful arms, a skeleton of stars, spiralling out from this core. The earth lives out in the sparsely populated arms orbiting the sun, an ordinary star, with our planetary companions. Our humble solar system. Here we are. A small planet, an ordinary star, a huge cosmos. But we're alive and we're sentient. Pooling our efforts and passing our secrets from generation to generation, we've lifted ourselves off this blue and green water-soaked rock to throw our vision far beyond the limitations of our eyes.

The universe is full of galaxies and their stars. Probably, hopefully, there is other life out there and background light and maybe some ripples in space. There are bright objects and dark objects. Things we can see and things we can't. Things we know about and things we don't. All of it. This glut of ingredients could carry on in every direction forever. Never ending. Just when you think you've seen the last of them, there's another galaxy and beyond that one another infinite number of galaxies. No infinity has ever been observed in nature. Nor is infinity tolerated in a scientific theory – except we keep assuming the universe itself is infinite.

It wouldn't be so bad if Einstein hadn't taught us better. And here the ideas collide so I'll just pour them out unfiltered. Space is not just an abstract notion but a mutable, evolving field. It can begin and end, be born and die. Space is curved, it is a geometry, and our experience of gravity, the pull of the earth and our orbit around the sun, is just a free fall along the curves in space. From this huge insight people realized the universe must be expanding. The space between the galaxies is actually stretching even if the galaxies themselves were otherwise to stay put. The universe is growing, ageing. And if it's expanding today, it must have been smaller once, in the sense that everything was once closer together, so close that everything was on top of each other, essentially in the same place, and before that it must not have been at all. The universe had a beginning. There was once nothing and now there is something. What sways me even more, if an ultimate theory of everything is found, a theory beyond Einstein's, then gravity and matter and energy are all ultimately different expressions of the same thing. We're all intrinsically of the same substance. The fabric of the universe is just a coherent weave from the same threads that make our bodies. How much more absurd it becomes to believe that the universe, space and time could possibly be infinite when all of us are finite.

So this is what I'll tell you about from beginning to end. I've squeezed down all the facts into dense paragraphs, like the preliminary squeeze of an accordion. The subsequent filled notes will be sustained in later letters. You could say this is the story of the universe's topology, the branch of mathematics that governs finite spaces and an aspect of space-time that Einstein overlooked. I don't know how this story will play itself out, but I'm curious to see how it goes. I'll try to tell you my reasons for believing the universe is finite, unpopular as they are in some scientific crowds, and why a few of us find ourselves at odds with the rest of our colleagues.

# 2

## INFINITY

14 SEPTEMBER 1998

I'm on the train back from London – gives me time to write, this time about Albert Einstein, hero worship, idolatry and topology. Somebody told me he is reported to have said, 'You know, I was no Einstein.' He couldn't get a job. His dad wrote letters to famous scientists begging them to hire his unemployed son. They didn't. The Russian mathematician Hermann Minkowski (1864–1909) actually called him a 'lazy dog'. Can you imagine? He worked a day job as a patent clerk and thought about physics maybe all the rest of his waking hours. Or maybe the freedom from the criticism of his colleagues just gave his mind the room it needed to wander and let the truth hidden there reveal itself. In any case, in the early 1900s he developed his theory of relativity and published in 1905 a series of papers of such import and on such varied topics that when he received the Nobel prize it wasn't even for relativity.

Now we love him and his crazy hair and he's considered a genius. We try to make him the president of a small country. He's a hero. And he deserves to be. When I think of his vision, his revolution, it's an overwhelming testament to the human character, one of those rare moments of pride in my species. Nonetheless, we've been led astray by our faith in Einstein and his theory. General relativity, as I'll get to later, is a theory of geometry but it is an incomplete theory. It tells us how space is curved locally, but it is not able to distinguish geometries with different global properties. The global shape and connectedness of space is the realm of topology. A smooth sphere and a sphere with a hole in the middle have different topologies and general relativity is unable to discern one from the other. Because of this, people have assumed that the universe is infinite – seemed simpler than assuming space had handles and holes.

*Assuming the universe is infinite is similar to assuming the earth is flat.*

*Explorers were feared to have sailed off the edge.*

*Instead they can sail the globe and end where they begin.*

*If the universe is finite, explorations of space may end where they began.*

Figure 2.1 *Is the universe infinite?*

I liken this to assuming the earth is flat. I suppose it's simpler, but nonetheless wrong. If you think about it, it's not so much that Europeans thought the earth was flat. They knew there were hills and valleys, local curves. What they really feared was that it was unconnected. So much so that they imagined their countrymen sailing off its dangling edge. The resolution is even simpler. The earth is neither flat nor unconnected. It is finite and without edge (Figure 2.1).

It's easy to make fun of an ancient cosmology, but any child will conjure up their own tale about the sky and its quilt of lights. I had my own personal childhood cosmology. I fully expected that the earth was round, but I got a bit confused thinking that we lived inside the sphere. If I walked far enough from our backyard, I was certain I'd hit the arch of the blue sky. For some reason I thought our backyard was closer to the edge of the earth. In my childhood theory, there is a clear middle point on the surface of the earth. The real earth is so much more elegant. The earth is curved and smoothly connected. There is no edge, no middle. Each point is equivalent to every other.[1]

---

[1] Of course each point on the earth is not identical to every other – Poland is not identical to Zimbabwe – but if the earth were a perfect sphere, each point would be equivalent to every other.

It is this and more that some cosmologists envisage for our entire universe: finite and edgeless, compact and connected. If we could tackle the cosmos in a spaceship, the way sailors crossed the globe, we might find ourselves back where we started.

Sometimes it's comforting, like defining a small and manageable neighbourhood as your domain out of the vast urban sprawl. But today the image sits uncomfortably. A prison thirty billion light years across.

Finally, the train's arrived. We're here. More soon.


**15 SEPTEMBER 1998**

Infinity is a demented concept. My mathematician collaborator scolded me for accusing infinity of being absurd. I think he'd be equally displeased with 'demented', but these are my letters, my diary. I only voice his objection for the record.

Infinity is a limit and is not a proper number. No matter how big a number you think of, I can add 1 to it and make it that much bigger. The number of numbers is infinite. I could never recite the infinite numbers, since I only have a finite lifetime. But I can imagine it as a hypothetical possibility, as the inevitable limit of a never-ending sequence. The limit goes the other way too, since I can consider the infinitely small, the infinitesimal. No matter how small you try to divide the number 1, I can divide it smaller still. While I could again imagine doing this forever, I can never do this in practice. But I can understand infinity abstractly and so accept it for what it is. Infinity has earned its own mathematical symbol: ∞.

All the greats have paid homage to the notion of infinity, each visiting the idea for a time and then abandoning the pursuit. Galileo Galilei (1564–1642) found the concept weird enough that he, like Aristotle (384 BC–322 BC) centuries before him, did not believe in the infinite outside of mathematics and maybe not even there. The Russian-born mathematician Georg Ferdinand Ludwig Phillip Cantor (1845–1918) was the genius who rigorously tried to understand infinity. He realized remarkably that there is more than one kind of infinity. There are an infinite number of them and surprisingly they actually come in different sizes. As John Barrow said at a lecture I gave in Cambridge, 'Some infinities are bigger than others.'

In Orwell's *1984* the doomed hero Winston Smith rebels against his oppressors with his defiant adherence to small numerical truths. Smith's sanity hinges on his commitment to one plus one equalling two and his rejection of one plus one equalling three. If Big Brother broke Smith's devotion and belief in arithmetic, he could break his mind. Cantor proved that infinity does not respect finite arithmetic and that in fact infinity plus infinity equals infinity. This would torment Winston's tormenters. Cantor was tormented. If not by his actual discoveries, then certainly by his singular obsessive attention to his spurned mathematics. He suffered a series of mental collapses but still managed to initiate a dramatic change in the direction of modern math. Cantor was able to develop an arithmetic that applied only to infinite numbers, *transfinite arithmetic*, an arithmetic so confounding and occasionally genuinely paradoxical that he himself is reported to have said, 'I see it but I do not believe it.'

I'll only make a brief excursion into transfinite arithmetic. It's enough just to follow the rhythm of it all, since it's poetic in a way; the melody of numbers. His arithmetic was based on aligning the elements of one set into one-to-one correspondence with the elements of another set. Consider the set of all natural numbers like $\{1, 2, 3, 4, \ldots\}$. The set of natural numbers is infinite – the series could go on forever. Half of the natural numbers are even, namely $\{2, 4, 6, 8, \ldots\}$. Every other natural number is odd, namely $\{1, 3, 5, 7, \ldots\}$. What Cantor realized is that there are nonetheless an infinite number of even numbers and an infinite number of odds. I can line the list of natural numbers above the list of even numbers and establish a one-to-one correspondence, so there must be the same number of elements in each set, namely an infinite number:

$$\begin{pmatrix} 1 & 2 & 3 & 4, & \cdot & \cdot & \cdot \\ \downarrow & \downarrow & \downarrow & \downarrow & \downarrow & \downarrow & \downarrow \\ 2 & 4 & 6 & 8, & \cdot & \cdot & \cdot \end{pmatrix}$$

This kind of infinity is countable. Surprisingly, we have to conclude that the set of even numbers is the same size as the set of natural numbers even though only half of the natural numbers are even. Infinity split into two produces two equal infinities: $\infty + \infty = \infty$. I'm abusing the proper mathematical notation here a bit. No one would write transfinite arithmetic this way since infinity really shouldn't be treated as a number but rather as the property of a set of numbers. The cardinality of countably infinite sets such as the natural numbers is written $\aleph_0$ and pronounced aleph-nought. What we really should say is that $\aleph_0 + \aleph_0 = \aleph_0$.

While the natural numbers and the even numbers are of the same infinity, not all infinities are created equal. Some infinities are bigger than others. The mystery of infinity is the legacy of the mystery of numbers. Starting with the ancient Pythagoreans, the set of natural numbers {1, 2, 3, 4, ...} was thought to be the core of mathematics, and by extension of reality. Imagine their dismay at discovering numbers that could not be built operationally out of the naturals. All numbers, it was believed, can be expressed as a simple ratio of natural numbers, such as one-half, which can be written as ½, the ratio of 1 to 2. The discovery of numbers that could not be expressed as a simple ratio of natural numbers struck fear in the hearts of the devout. This set of irrational numbers, as they are now called, was first discovered by the Pythagoreans through the study of geometry.

The Pythagoreans were so seduced by the sanctity of natural numbers that the discovery of irrationals cast a gloomy cloud over their faith. The famed 'Pi', written $\pi$, is irrational and has deep geometric meaning. The circumference of a circle equals $\pi$ times twice the radius (Figure 2.2). The numerical value is $\pi = 3.14159265\ldots$, where the series following the decimal point never ends. The Chinese found an excellent rational approximation, $\pi \approx {}^{355}/_{113}$. But this is only an approximation and eventually the true number that is $\pi$ differs from the truncated list of numbers that is 355 divided by 113. The golden mean is another irrational derived from simple geometry[2] and is equal to 1.618033989... The golden mean is the ratio that results when a segment is divided so that the ratio of the longer to the whole is the same as the ratio of the shorter to the longer (Figure 2.3).

All of the numbers between any two natural numbers fall into one of

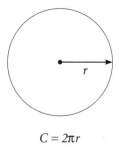

$$C = 2\pi r$$

Figure 2.2 *The irrational number $\pi$ is nested in geometry. The circumference of a circle equals twice the radius of the circle times $\pi$.*

---

[2] The golden mean can be written as $(1 + \sqrt{5})/2$ or as the infinite sequence of ratios: $1 + \cfrac{1}{1 + \cfrac{1}{1 + \cfrac{1}{1 + \ldots}}}$ known as a continued fractional expansion and abbreviated as [{1}].

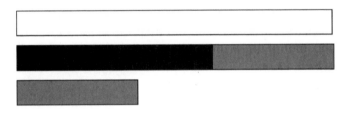

Figure 2.3 *The golden mean is the ratio of the black segment to the white and also of the grey segment to the black.*

these two sets, either rational or irrational numbers. There are an infinite number of numbers between 0 and 1. Nature packs infinity in the most humble interval. The rational numbers falling between 0 and 1 can be represented as one integer divided by another, like ½. There are an infinite number of these and the set looks like {½, ⅓, ¼, ...}. Since the rational numbers are just one natural number divided by all of the other natural numbers, the set of rational numbers is the same size as the set of natural numbers.[3] An infinite set of this size is said to be countably infinite. The irrationals are not so simple.

---

[3] This is actually not immediately obvious. Slowing down a bit, the collection of all fractions can be written as the collection of lists

$$\{1/1, 1/2, 1/3, 1/4, ...\}$$
$$\{2/1, 2/2, 2/3, 2/4, ...\}$$
$$\{3/1, 3/2, 3/3, 3/4, ...\}$$
and so on.

Each row above is generated by a natural number. The set of rows is therefore the same size as the set of natural numbers and must have cardinality $\aleph_0$. Since each row then contributes an additional infinite sequence to the counting, it seems that the set of all fractions must be larger than the set of natural numbers. However, this is not the case. Since the unit fractions in the first row, $\{1/1, 1/2, 1/3, 1/4, ...\}$, are just 1 divided by all of the other natural numbers, the set of unit fractions automatically aligns into one-to-one correspondence with all of the natural numbers and therefore is a set of the same size, a set with cardinality $\aleph_0$. The next row can be written as 2 divided by all of the natural numbers and so also has cardinality $\aleph_0$. Each separate row has cardinality $\aleph_0$ and the set of such rows has cardinality $\aleph_0$. So the set of all fractions must have cardinality $\aleph_0 \times \aleph_0$. In a finite arithmetic, this would have to be larger than the set of natural numbers. But infinite sets do not obey finite arithmetic. While it is a bit tricky, all fractions can then be re-ordered to show that there does exist a one-to-one correspondence with the natural numbers and therefore in transfinite arithmetic $\aleph_0 \times \aleph_0 = \aleph_0$. What one shows is that the set can be systematically counted and is therefore countably infinite.

Cantor realized that the set of irrationals was infinite in a way that was so huge as to be uncountable. They cannot be represented as one integer divided by another and some require an infinitely long description, such as $\pi$ or the golden ratio. An uncountable infinity could never be put into one-to-one correspondence with a countable infinity and so the irrational numbers must be of a larger infinity than the natural numbers.

Cantor proceeded to define the continuum as all of the numbers between any two natural numbers. So the interval between 0 and 1 is comprised of a continuum of numbers, a countably infinite number of rationals and an uncountably infinite number of irrationals. The entire continuum is an uncountably infinite set. Pulling infinite sets out from between the seemingly benign interval from 0 to 1 feels a bit like pulling an elephant from a box. He eventually realized that there was an infinite hierarchy of infinities. It reminds me of a Thomas Hobbes quote, 'To understand this for sense it is not required that a man should be a geometrician or a logician, but that he should be mad.'

Even mathematicians rejected both the notion of infinity and Cantor. Like every genius before him and since, he encountered violent opposition. Very clever and influential people squashed Cantor's vision. The most forceful was Leopold Kronecker (1823–1891), who had his own mathematics based only on the finite numbers which rejected even negative numbers. I don't see the harm in negative numbers, but I suppose he had his reasons for their exclusion. The nastiness between Cantor and Kronecker some blame for simultaneously wounding and shaping modern mathematics. Cantor was no doubt personally wounded by this rejection and is known to have grappled with profound depression. He worked on the periphery of mainstream mathematics and passed many bouts of mental illness committed in institutions, and I don't mean academic ones. It wasn't a battle he won. He would die there. Not to be bleak. Cantor would never live to see the powerful impact his theories would have on twentieth-century mathematics. A tragic theme that keeps repeating.

**9 OCTOBER 1998**

We finally found a flat in Brighton. We spent a harsh month commuting from London, but now we're here and feel like we can survive anything if we could survive this move. Our bad spell culminated in a minor explosion. Yesterday Warren blew up the computer, plugging the 120-Volt transformer into the 250-Volt British outlet. We couldn't believe our eyes when the spark flashed and the machine started to smoulder. So we're feeling a bit sorry for ourselves, but we're getting over it with visits to the Brighton pier at night. The carnival music tries to fuel a festive spirit in the few couples bundled against the ocean air. We ignore the abandoned rides or marvel at them and their loneliness. We stand on the pebble beach and try to throw rocks to America against the wind.

Despite feeling displaced, or maybe because of it, I am getting settled at work. I'm a bit excited about some results my collaborators and I accumulated in Berkeley. The Berkeley team has scattered. Joe Silk is in Paris. I'm here in England with John Barrow. Only Evan Scannapieco is still in California, finishing his PhD. Giancarlo de Gasperis is back in Italy to finish his degree. I saw Giancarlo in Rome last month. We drank strong coffees standing at a bar after a conference. The Roman background suited him.

It won't be meaningful to you yet, but just to lend a visual, while at Berkeley we modelled a sphere with bright spots, tracing symmetries around the surface with a recurrence of the number five, a five-pointed star and five-sided polygons (Figure 2.4). The pattern of spots is our conjecture for what the sky might look like to future satellite missions if the universe is small. I'll get to this later. Before I reject infinity, I want to admire it.

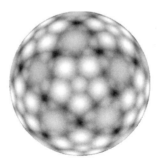

Figure 2.4 *A possible pattern that could be encoded in the light left over from the big bang.*

There have been different ideas on what is real in mathematics and what is invented. Kronecker didn't believe in negatives, the Pythagoreans were frightened by irrationals, almost everybody but Cantor abandoned infinity. John Barrow was entertaining me with other anecdotes in the war of the mathematicians that saw more fits of madness and bitter rivalry. Many of these stories are in his book *Pi in the Sky*, but he's run out of copies to give me. I pretended to be incredulous when he suggested it could be found in the local book store. John has spoiled me with donations of his many other books to my library. In John's book I found this outrageous quote attributed to the Dutch mathematician Luitzen Egbbertus Jan Brouwer (1881–1966): 'All my life's work has been wrestled from me and I am left in fear, shame and mistrust, and suffering the torture of my baiting torturers.' Brouwer was perhaps a more moderate adherent to the ideas of Kronecker, but moderate in temperament he was not. He is reputed to have been a pessimistic, outspoken misogynist, and those are among the nicer things people have to say about him. I suppose he was a true misanthrope, so his venom was not solely aimed at women, but it was particularly acerbic when so directed. He also suffered from nervous attacks if not full-blown mental illness.

Brouwer was in the camp which held that mathematics was discovered in the physical world and born of an experiential intuition. Like Kronecker, he rejected infinity in mathematics because he didn't find it in nature. They both believed all of mathematics could be derived from the natural numbers. Then there were those like Cantor who believed that if a concept such as infinity was logically self-consistent then it was sound mathematics even if it could not readily be found in nature. It seems a semantical distraction to argue on the existence of Cantor's mathematics. Certainly it exists, if only in our minds. Isn't it real enough even if it exists only in the configuration of our thoughts? We can't be intimidated into ignoring these thoughts. One plus one may equal two, but $\aleph_0$ plus $\aleph_0$ equals $\aleph_0$. Cantor created a world of infinities for us to play with and I'm on his side and glad his mathematics survived and triumphed. But I don't know if infinity has a place in nature.

There's a good paradox due to Zeno (about 490 BC–about 425 BC), the ancient Greek philosopher from Elea, which is now southern Italy. Learned in the pre-Socratic schools of Greek philosophy, he is believed to have written an influential but now lost book on infinity. He was mystified by the idea of a continuous series known as the continuum and argued that if any given distance could be divided in half then the two resultant pieces could be divided in half. Repeating the process an infinite number of times, there must be an infinite number of pieces across even an inch. We could never cross the room because we would have to pass an infinite number of points before reaching the other side. We would have to move past the smallest infinitesimal piece infinitely quickly. The argument suggests we shouldn't be able to move any distance at all and so motion itself should be impossible. Yet we do move.

We make it across the room without a thought for the pilgrimage across a landscape of infinity that simple motion involves. While infinity is an elegant and important idea in mathematics, it is shunned by the physical. I don't know of any simple resolution to Zeno's paradox, but I can still move.

The idea that we can all be broken down to fundamental indivisible quanta, bundles of energy and matter, might save us from Zeno's paradox and get us to move across the room. An inch-long ruler cannot be divided an infinite number of times because eventually it will be reduced to its fundamental quantum particles, which are themselves indivisible. In other words, there is no reality to the physical continuum because all physicality comes fundamentally in discrete, quantized units. Motion across the room is permitted because only a finite number of quanta need be passed, whether we know it or not. Even this doesn't really spare us confusion, since Zeno had at least forty such paradoxes, some of which quanta can't escape either.

Don't get me wrong, I don't believe that math and nature respond to democracy. Just because very clever people have rejected the role of the infinite, their collective opinions, however weighty, won't persuade mother nature to alter her ways. Nature is never wrong. Still, I don't believe in the physically infinite.

Where in the hierarchy of infinity would an infinite universe lie? An infinite universe can host an infinite amount of stuff and an infinite number of events. An infinite number of planets. An infinite number of people on those planets. Surely there must be another planet so very nearly like the earth as to be indistinguishable, in fact an infinite number of them, each with a variety of inhabitants, an infinite number of which must be infinitely close to this set of inhabitants. Another you, another me. Or there'd be another you out there with a slightly different life and a different set of siblings, parents, offspring. This is hard to believe. Is it arrogance or logic that makes me believe this is wrong? There's just one me, one you. The universe cannot be infinite.

Of course, my faith in nature and its laws is deeper than my need for uniqueness. If I truly believed there was no way for the laws of physics to be consistent with a finite universe, I might be swayed. But there are ways, simple ways, for the laws of physics to be consistent with a finite universe. The universe can be created a finite size in the big bang. I didn't realize this until long after I had finished my doctorate. The subject of topology, the global shape of space, is not really taught to us. But in a way that's good because it gave me something to do as a postdoc. First we have to learn about general relativity and before that special relativity, which means we can talk about Albert Einstein. And before him Newton.

So for the record, I welcome the infinite in mathematics, where my collaborator is right: it is not absurd nor demented. But I'd be pretty shaken to find the infinite in nature. I don't feel robbed living my days in the physical with its tender admission of the finite. I still get to live with the infinite possibilities of mathematics, if only in my head.

# 3

## NEWTON, 300 YEARS AND EINSTEIN

**26 OCTOBER 1998**
From the window up here on the second floor, I can see Warren clomping down the street. He makes frequent trips to the decrepit grocery store across the road for food so cheap you wonder how they can make it for that. Sadly, the food is worth the price and I'd happily pay a large fraction of my salary for a form of food without mayonnaise. (I've seen mayonnaise in the sushi.) It's the small things that induce culture shock.

Just so you don't think I'm disparaging my new home, they have been lovely to me here, welcoming and encouraging. I feel inspired to work and though pressured by the strains (they say moving is more stressful than death or divorce) I think maybe this was the right move. Besides I'm a shock to the culture. I stand up to give a talk and find myself hesitating slightly before letting the heavy American sounds ring out. I feel conspicuous. I try to focus on phonetics, the weight of the consonants and the relative thudding of the American 't'. What would Newton have made of this? He didn't even want Catholics in his college let alone Yankees (anachronism).

Here we begin with Sir Isaac Newton (1643–1727). An eccentric genius. Newton's theory of gravity describes an apple's fall from a tree and the earth's orbit around the sun. Or maybe we should go further back, back to Nicolaus Copernicus (1473–1601), who argued for a cosmic humility where we were not at the centre of the solar system. Watching the inevitable rise of the sun and night relentlessly follow day, it might be no surprise that civilizations used to believe the sun orbited us. In the Copernican view, the sun does not orbit the earth but we it, along with a collection of other planetary rocks (Figure 3.1). While this idea may have long predated him, it was really Copernicus who

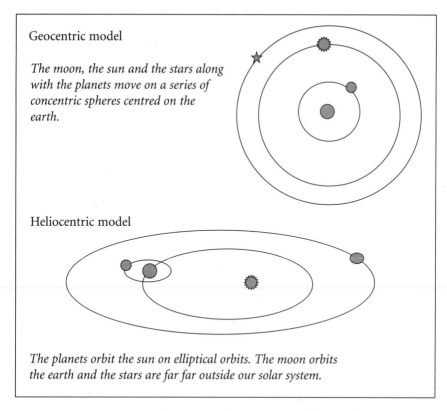

Geocentric model

*The moon, the sun and the stars along with the planets move on a series of concentric spheres centred on the earth.*

Heliocentric model

*The planets orbit the sun on elliptical orbits. The moon orbits the earth and the stars are far far outside our solar system.*

Figure 3.1 *The geocentric model versus the heliocentric model. The Copernican model with the central sun describes our solar system.*

dethroned a persistent hubris that we were special, actually central to the cosmos. His ideas were strongly resisted by other philosophical principles, particularly religious tenets. The European nobleman and astronomer Tycho Brahe (1546–1601) took it upon himself to propose an observational resolution. It was his suggestion that we try to measure the motions of the heavens and thereby observe which was right, the heliocentric or geocentric model.

Tycho Brahe had lost his nose and died of some combination of politeness and gluttony. Engorged at a banquet, he managed to die of a burst bladder on his carriage ride home. At least that's as I remember the story. The nose he lost in one of many duels fuelled by a reputedly surly personality. He is rumoured to have worn a gold surrogate in its place.

Tycho was able to build an astronomical observatory on an island off Denmark, equipped with elaborate facilities. He veritably ruled the island, empowered by the nepotism of nobility. The observatory allegedly had a gaol, or maybe it has more medieval flair to say dungeon. I won't speculate on the crimes of the interned. This peculiar little man collected a vast amount of data, charting the skies from Denmark and later in exile from Prague after abrading his patron. Remarkably, his observations were performed without telescopes, a technological advance to come only after his death. In Prague he worked with an assistant, Johannes Kepler (1571–1630), a mathematician and Tycho's famed successor. Amusingly, Tycho decided Copernicus was wrong, although we now know that Copernicus was right. All the other planets were the sun's satellites but the sun itself orbited us, or so he argued. If the earth were in motion around the sun, he reasoned, we would see the other objects in the heavens move. He found no such parallax, although today we can make sensitive measurements confirming the relative motion of the stars due to the earth's orbit about the sun. He would have been forced to conclude that the stars were shockingly huge and distant to elude observation of their motions. Fact is stranger than fiction. The stars are huge, thousands of times the size of the earth and very far away. The solar system is just our backyard. The world is huge. The universe is huge.

Kepler, on the other hand, was a proponent of the Copernican solar system. After Brahe's death in 1601, Kepler spent over twenty years pouring over Tycho's surplus of observations. Kepler was able to deduce three laws that beautifully describe the motion of the planets around the sun. The earth-centred model truly lost and the solar system model survives. Kepler's three laws:

(1) The planets move on closed ellipses around the sun. In Einstein's theory, this isn't true. The first confirmation of general relativity was that Mercury's orbit does not trace out a closed ellipse, but instead the orbit follows a precessing ellipse: that is, the direction of the ellipse drifts slowly around the central sun (Figure 3.2). Einstein has loomed large in my life and we will inevitably come to his trials and triumphs.

(2) A given planet moves fastest on closest approach to the sun and slowest at the point of farthest approach. The motion is such that the orbit sweeps out an equal area of the ellipse for equal times.

(3) His third law relates the period of the orbit to the size of the orbit and would later be derived as a consequence of Newton's theory of gravitation.

Galileo meanwhile was building a theory of dynamics, a theory of the nature of motions, as Kepler's empirical laws were formed. Galileo discovered the idea of inertia, which has worked its way into our

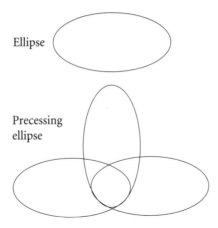

Ellipse

Precessing
ellipse

Figure 3.2 *Top: A closed ellipse. Bottom: A few stages in the precession of an ellipse.*

vocabulary as a social notion. The principle of inertia asserts that objects will travel in a straight line unless acted upon by a force. In the absence of forces, objects will not come to rest but will instead move with constant velocity. It's a metaphor for life. Galileo Galilei unknowingly passed the burden of the history of gravitation to Isaac Newton, born in 1642, the year of Galileo's death. The idea of inertia was refined profoundly by Newton and ultimately led Einstein some 300 years later to propose the principle of relativity. And here we are back to Newton, which is where I wanted to start in the first place.

## 5 NOVEMBER 1998

During our month of wandering around the United Kingdom we intended to have fun and failed. Finding our flat was an ordeal and I won't bore you with our tales of misadventure. I can't help but remember the bedsit we found in Brighton as an act of desperation to end our wanderings. Electricity in the bedsit was coin operated. You ran out of coins, you ran out of light. I had always heard of such things in the old world, but in all my travels this was my first coin-op bedsit. I was feeling robust enough to be amused. Warren, on the other hand, sat on the edge of the bed catatonic, staring at the woodchip wallpaper. He confided in me later that it was the woodchip which disturbed him most, an odd cue for the memories of his Manchester childhood. I was overjoyed to see that bedsit. It was going to be our private home for nearly two weeks, which seemed a long time in context. We had two pans, an electric burner, and a fire alarm placed inconveniently close. The alarm and cooker worked as a team and often sent us into a panic during even the tamest cooking ventures, toast being most troublesome.

Now finally we have our own flat and the blue mood is lifting. We are edging closer to Newton's home and may even move there, although we just got here. What a transition. My mind has ridiculously linked bedsits, toast and Newton. Newton developed his ideas both as a student and as a professor at Cambridge, continuing to devise his theories even as the plague ravaged Europe and forced a year long closure of the university. His ideas were set forth in the entombed and fondly abbreviated *Principia* from the full Latin name *Philosophiae Naturalis Principia Mathematica*. At the colleges in Cambridge and Oxford they still make the Fellows use Latin in ceremonies. Apparently one of my Yankee friends roguishly showed up in jeans without his academic gown, which apparently he doesn't own anyway, and not speaking a word of Latin. But he did know that in *Principia* the 'c' is pronounced like a 'k', which is a start. I believe he said in his defence, 'I'm just from one of the colonies.'

Newton suggested that gravitational mass, which is related to the weight of a body under the earth's pull, and inertial mass, which is related to the resistance of a body to motion, were one and the same, a subtle prediction that has survived experimental tests. He elevated this notion to a universal principle, suggesting that all masses pulled all other masses and that the strength of this pull grew weaker with distance. Using Kepler's observationally determined laws and Galileo's theories of motion, Newton was able to construct a mathematical expression for the force of gravity. In the *Principia* he presents three physical principles drawn from Kepler's observational laws:

(1) Every body moves in a straight line at constant velocity or remains at rest, unless acted upon by a force.
(2) The direction and the magnitude of the change in the motion is proportional to the force.
(3) To every action there is an equal and opposite reaction. Another metaphor for life.

In the process Newton invented calculus, the mathematics essential to modern physics. With calculus we can understand in equations how dynamic systems evolve with time. This formalizes determinism. Put in an initial condition and we can follow the equations to an inevitable, precise outcome. The universal nature of Newton's insights entrenched the notion of determinism in natural philosophy. The deterministic nature of cause and effect became central to other branches of philosophy and has had obvious influence on our modern cultural outlook. Determinism and causality are weakened by quantum theory, which though poorly understood nonetheless works. And don't get me started on quantum mechanics and determinism. There is an unresolvable philosophical debate that recurrently rears its ugly head on the impossibility of free will in a life dominated by determinism. The distilled and simplified argument goes something like this: if every atom in our bodies merely follows a mechanical trajectory precisely determined by the laws of physics then we have no volition. Our choices are predetermined and we merely play out the inevitable effect of all those earlier causes.

A deterministic universe is like a movie where the end is already recorded. We don't know the ending, so we have the impression that it's unfolding in real time and a sense of spontaneity, but the end *is* already written, already determined. Maybe nature has restricted our perception in this way to protect us from the completely bleak state of affairs of knowing the ending, but it's an illusion all the same.

People used to try to hijack quantum mechanics and its inherent mystery to cast a cloud around determinism, in the hope that free will could survive modern physics. But that never worked very well. Since when does random chance equal free will? The only salvation for volition is a soul and faith and you're not allowed to ask me about that.

A thread from Copernicus to Tycho to Kepler to Galileo to Newton wove this picture which successfully predicted the motion of the planets and the moon. Incidentally, if anyone wants to deconstruct the history of science, I have no objections. Who knows who else participated and here I go perpetuating Eurocentricism and other politically malicious notions. But I'm no historian and it makes for a nice tale.

There were 300 years between Einstein and Newton. Those 300 years were dominated by Newtonian ideas. One bit of information that did not sit well with Newton's model was a peculiarity in Mercury's orbit. The perihelion of the orbit, the point of closest approach, was observed to precess: that is, the elliptical trajectory drifts around the sun, while Newton's and Kepler's laws predicted that the elliptical orbit should be perfectly closed without precession. The precession of the perihelion of Mercury was the first observation to confirm Einstein's relativistic theory of gravity. Einstein may or may not have had this bit of evidence in mind when he pursued a revision of Newton's laws. It is often said that Einstein's motives were more philosophical.

In the couple of decades before Einstein's college years, the Scottish scientist James Clerk Maxwell (1831–1879) developed a remarkable unification of electricity and magnetism in one elegant theory. Modern electromagnetism and Newtonian theory did not fit together perfectly and the first hints that something deep was at work began to trickle in. Here comes Albert. It seems to happen in the course of scientific history that two superb theories will clash and one of the edifices, if not both, will have to give. The ultimate successor always ushers in a new era of thought and is never short of a revolution. Einstein incited a revolution, a revolution that managed to preserve Newtonian ideas where they lay claim to our intuition but yielded to a theory of relativity in the more extreme realms beyond our everyday experience.

What I really want to tell you about now is special relativity. That's where we're going. Einstein had two theories of relativity. The first came in 1905 – that was special relativity. Then some years later in 1915 he really outdid himself with the theory of general relativity. General relativity is a theory of gravitation and curved space. It seems that no matter what I work on, black holes, chaos, the big bang, the one theme in common is always general relativity. That theory is the generator or inspiration for all of these phenomena. So I'm really a devotee. I'm hooked. We'll talk about gravity later. Let's start with special relativity, which intentionally ignores the influence of all forces including gravity.

# 4

## SPECIAL RELATIVITY

**3 DECEMBER 1998**

I live with an obsessive-compulsive maniac musician from Manchester. His behavioural disorder is oddly endearing and he does his best to keep it to himself. He still seems shocked that I caught him tracing triangles or that I could identify the shape when he tried to covertly trace their three sides on my back. He traces triangles everywhere, counting the corners in his head 1-2-3-4-3-2-1. No pause. A smooth transition from 3 to 4 and back to 3 again. Cheating 4. The three others each appearing twice for the one appearance of 4. Its 4's fault. 4 shouldn't be there at all. Only three corners on a triangle. What's that 4 for anyway?

Triangles trace him too. His hands and feet are tapered into triangles. He has the usual collection of two hands and two feet. Making four triangles on his form. That must be what the 4 is for. I asked him if his fingers moved independently of each other or if they always coordinated as a unit. He stared at his tapered hands, delighted with me for noticing and with himself for harbouring such unique appendages. The theme of triangles fills his head with contentless numerical patterns, but he fends them off by rehearsing an old-time tune with the attention that only an obsessive could deliver.

I myself have been less than even-keeled. I've been working with the fury of a mad woman. I've been so wired up, heart pounding, I could feel the pulse swelling in my neck. But I was productive while the mania lasted. Now I'm a bag of protoplasm, waiting for the next rush of adrenaline. I'm slouched on the couch, which is less the fault of my posture than of the cheap, worn stuffing characteristic of this rental flat. I readjust to hang off the edge and watch Warren. He's maybe not what I expected, a rough boy with a voice like chocolate and a penchant for country music. But I love the blond around his sideburns and the funny way he walks. He's brave coming with me back to England. We're in this together. My partner in crime.

Some very clever people were obsessive-compulsive. I don't believe insanity is either a requirement or a guarantee for brilliance. But I find the anecdotes so interesting, so much more interesting than the usual hero worship I'm subjected to by my brothers in science. Sharon Traweek wrote a sociological study of particle physicists in a book called *Beamtimes and Lifetimes: The World of High Energy Physicists* with at least one chapter on this annoying habit of iconofying the men of science. I find their weaknesses so much more touching.

Newton wasn't obsessive-compulsive to my knowledge, but the tenacity of his mental health has certainly been called into question, particularly in his later years. Newton was a secret alchemist, conducting covert experiments in his college rooms in Cambridge, including very peculiar ones that involved staring at the sun and stabbing himself in the eye with a small dagger. His mental ailments are usually described as paranoia and depression. Some have even suggested that he was as mad as a hatter, meaning his insanity was induced by mercury and other chemicals he ingested in the course of his alchemy – chemicals that led to the mental disintegration of traditional hatmakers. Others suggest his emotional breakdowns were incited by the trials of his covert homosexuality. A broken heart, that sounds more likely.

Any mental lapses seem to have had little impact on his intense scientific clarity, at least for most of his production. Newton was so right about so many things that it seems ungenerous to dwell on where he was wrong. It was really Newton who first proposed a principle of relativity. He argued that the laws of physics should be the same for all observers moving uniformly in the absence of forces: that is, all inertial observers. That was a very strong intuition that Einstein took to heart. When it came to mechanics and Newton's laws, the equations did appear the same for all observers, those moving and those at rest. Consequently, when you're on a plane and there's no turbulence you can drink coffee, put your book down, walk around. Things behave mechanically just as you expect them to when you're stationary on earth. Last time Warren and I flew back to the States we were bumped up to first class. It was more comfortable than our TVless Brighton flat. It was our first dinner out in months. We had a white tablecloth and beautiful cheeses, movies and mimosas. They even offered a massage. We were clutching hands in a state of blissful enthusiasm. They never had such grateful passengers.

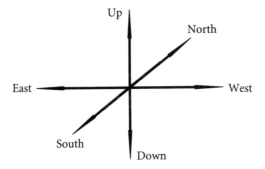

**Figure 4.1** *Three spatial dimensions: up–down, north–south, east–west.*

Despite the enhanced comforts, life operates as normal as long as the plane moves smoothly. Simple mechanics, as described by Newton, is the same for everyone moving smoothly, which is a good thing. The difficulty came in defining states of rest versus states of motion. And here Newton erred. He argued that space and time were absolute, that they defined a rigid coordinate system with respect to which we could unambiguously determine if we were moving or not.

Space is the three physical dimensions we freely occupy. The dimensions define the corner of a cube: there is up and down, north and south, east and west (Figure 4.1). Three spatial dimensions in all. According to Newtonian ideas, we all experience space the same, regardless of how we move through it or where we are. Time is a distinct coordinate and, according to Newton, is also absolute. If time is absolute, we will all experience the same flow of time, age the same, and watch the cycle of growth and decay with the same rate of change, regardless of our motion and location. Although our daily experience convincingly confirms this intuition, it is not true. Different observers measure space differently and experience a different passage of time.

If space and time were absolute then we should be able to identify absolute space when we look beyond the mechanics of rocks, planes and trains and consider theories like the theory of electromagnetism. James Clerk Maxwell gave insight into the nature of electricity and magnetism as one force. In Maxwell's unified theory of electromagnetism, light is an electromagnetic field oscillating like a wave. Sometimes it is a great advantage to bury your mind in the formalism of the equations because we often understand math when plain English simply isn't as useful. We don't often worry about the real meaning of a concept such as 'field'. As long as we can write down an expression, we can calculate, predict and string the modern world together with points of light and radios and

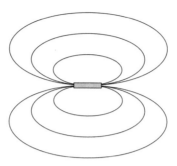

Figure 4.2 *Magnetic field lines.*

phone lines. In the absence of mathematical formalism, there is a canon-
ical illustration of a field which makes the concept more tangible. If I
tossed magnetic shavings in the presence of a magnetic field, the shavings would gather along the field lines, showing the presence, direction
and shape of an unseen 'field', for lack of a better word (Figure 4.2). This
field has energy and is as real as particles.

The discovery that light is a wave led people to ask the obvious question: a wave in what? Water waves are formed by the group motions of
water molecules, sound waves by the group motions of air molecules,
drum beats are waves in the taut skin of the drum. Light is not of this
kind. It was originally, and mistakenly, thought that light needed to
move through some medium. The existence of an aether was proposed
as the medium through which light could wave but no aether exists.

This was the first real crisis faced by Newtonian mechanics. If an
aether exists, it must be at rest with respect to absolute space. Since the
earth orbits the sun, we must not be at rest with respect to the aether and
absolute space. If Newton was right, then by charting the earth's motion
through the invisible aether we could identify the state of absolute rest.
One way to measure our relative motion is to measure the speed of light
as the earth orbits the sun. The speed of light should increase in one
direction and decrease in the other, just like a train approaches faster if
you run towards it than if you run away. Michelson and Morley devised
experiments to accurately measure the speed of light and discovered that,
contrary to the Newtonian prediction, the speed of light was exactly the
same in all directions and at all points along the earth's orbit. It always
moves and always at the same speed, which everyone denotes with the
simple symbol $c$. The speed $c$ equals 300,000 km per second, which is
pretty fast. It takes light only a fraction of a second to cross a continent
and only about 16 minutes to make it all the way to the sun and back.

Something had to give. If space and time were absolute, the speed of light must be relative and must depend on the observer's speed. But the speed of light refuses to depend on the observer's speed. The absolute structure of space and time and the relative speed of light switched places in Einstein's theory of special relativity. In special relativity, and in reality, as far as we can assess it, the speed of light is an absolute constant and the structure of space and time is relative.

Before Einstein there were a band of clever people sharing ideas who, somewhat tentatively, suggested that neither length nor time was absolute. By permitting space to contract and time to dilate they were able to cast Maxwell's laws into a form that appeared the same for all observers in the absence of forces. This is a significant philosophy to live by. What they said is that the laws of physics, being the same to all inertial observers, are a more important guiding principle than the absolute nature of space and time.

It was really Einstein who stood strong when he came to the idea that there was no absolute time and no absolute space. We cannot measure our velocity relative to absolute space because absolute space doesn't exist. The laws of physics must therefore be the same for all observers in relative motion, since these observers are truly equivalent. Light requires no medium, no aether. It is pure energy and propels itself forward as the oscillation of an electromagnetic field.

A light wave is a specific configuration of oscillating electromagnetic fields. Maxwell's laws determine the speed of light, and if Maxwell's laws are to look the same to all observers in the absence of forces, then the speed of light must look the same.

Einstein allegedly had a few inspired conversations with a friend on the subject and then, after taking some time to himself, barged in to declare: there is no aether, no absolute time, no absolute space. How thrilled he must have been to have seen that far.

I think about Newton saying, 'If I have seen farther, it is because I have stood on the shoulders of giants.' (I have to confess it was my neighbour, a manufacturer of ladies' knickers, who correctly identified the originator of that quote. When I tried to place the author of the adage, he said to me over tea, 'Speaking as an MA in frock design, I think it was Newton.' He was right: Newton is generally credited with this quote, although my editor tells me the actual history of the sentiment may be quite complicated.) How lucky we are to be able to clamber to their height, and maybe they hold us up and let us see just beyond their own view. Sometimes this is an arduous and unpleasant climb, but when you see things fall into place and nature stuns you with her harmony, it's enough to make you grateful.

## 5 DECEMBER 1998

I love the way Einstein thought. He had what I imagine to have been a very rich inner life. Most scientists are obsessed with experiments and look to observations to direct their ideas, which is fair enough. People tend to create a kind of phenomenology to explain the apparent facts of life. Einstein does seem to have operated differently. He invented thought experiments when actual physical experiments were impossible. A thought experiment is purely hypothetical – an invented game with strict, simple rules. The execution of these thought experiments helped him peer at the essence of space and of time.

He equipped imaginary observers with a system of rulers and clocks. He suspended any beliefs he may have held on the meaning of space and time, beliefs so ingrained that they froze other minds. He accepted that space is nothing more than the length according to the ruler of a given observer and time is nothing more than the ticks read on a clock.

The observers conjured up in his thoughts would perform a variety of experiments, diligently measuring distances and times between events. He would carefully examine how space travellers moving at near light's speed would communicate with people back home. He would carefully compare the readings on the different sets of clocks and rulers. What he found is that observers in relative motion would not measure the same distance between events and would not experience the same passage of time. He managed to demystify the notion that these properties could be relative.

He separated twins, drove cars through barns and launched rockets in a fantasy world made no less fantastic by its adherence to logic. With these imaginings, carried out in his truly unique mind, he overturned all of the familiar ideas of simple Newtonian mechanics.

This is what he came up with, two fundamental precepts under-pinning special relativity from which everything else can be derived: (1) the principle of relativity and (2) the constancy of the speed of light. Over the weeks following his revelation that there is no absolute time and no absolute space, Einstein derived the consequences of these two precepts. He eked out all of the famed results including $E = mc^2$. He discovered that time dilates, space shrinks and mass grows as we near the speed of light.

## 8 DECEMBER 1998

Brighton. We live off an old alley across from a pub called the Queen's Head, which dons a picture of Freddy Mercury's head from the operatic rockers Queen. We're a stone's throw from the beach and I think it's what I will remember of this time. It will be a physical memory of the cool smell of the ocean and the wind and light rain and the sadness I'm fleeing when I run along the promenade. I know we won't stay here long. A year at most. We can move to Cambridge where I could work in the maths department, or I could take a faculty job chosen from the few lectureship offers that have started to trickle in. Every night Warren or I call a pow-wow. The two of us sit down somewhere, in a pub or a coffee shop, or on the pebbles of the beach.We draw flow charts and diagrams. We can live in London and commute, one of us or both. We can live in Brighton so he can record with the fiddle player he met and I can commute. I can turn down all the jobs and we can go back to America, a land he loves.

He is weighed down by his memory of childhood in England, his Manchester home, the memory he'd hoped he'd forgotten but now burdens him. He wants to escape. Every day the plan changes and gets more intricate. It's up to me in the end. No matter how much I try to include him in the decision, we both know it's up to me. It's my work we're following. It's not all gloom. There are moments of real inspiration and we laugh our way through most of the crises.

He never asks me about my research. It's a relief. I come home and we fall into a linked privacy. We're together in our solitary thoughts. He studies music and I study math. We share curiosity, if not the object of interest. He thinks about bluegrass and today I think about Einstein.

Einstein's principle of relativity reinstated Newton's intuition with unexpected consequences. The principle asserts that the laws of physics must be the same for all inertial observers, for all observers moving freely in the absence of forces. All observers will experience the same laws of mechanics, measure the same speed of light, experience the same consequences of atomic interactions. All that matters is relative motion. No observer could ever prove it is the other that is moving.

The constancy of the speed of light in conjunction with the principle of relativity forces two observers in relative motion to disagree on their measures of space and time. Einstein would ride his bicycle and watch the light catch on the leaves and then sneak through to speckle the ground. He rode and wondered what it would be like to move as fast as a light beam. If Einstein could outrace a light beam, light would appear to stand still. But light can never stand still. It always travels at light's speed. Neither Einstein nor anybody else could catch up to a light beam.

If you run towards a light beam, its speed is *c*. If you run away, its speed is still *c*. Since speed is by definition a distance per unit of time, the observer running towards the light beam and the one running away must measure different distances and times in order to measure the same speed of light. Einstein showed that time must dilate and space must contract relative to any other set of observers armed with synchronized clocks and rulers. The time dilation means that clocks literally appear to tick slower. All clocks appear to run slower, including our biological clocks. The space contraction means that rulers would literally appear shrunken. All distances appear contracted, including the length of a room or of an outstretched arm.

Since motion is relative, it is impossible to determine who is really moving. It is meaningless to ask or answer this question. Each observer will see the other's time dilate and rulers contract. If I moved at nearly the speed of light past you, you would see me talk slower, my clock run slower, my heart beat slower. I would see you talk slower, your heart beat slower, your clock run slower. I would look all squeezed to you, but you would look all squeezed to me. Which is right? Which is true? Both are true. There is no objective answer to whose clock actually runs slower or whose face is truly squashed. I have no impression of time running slower or of space contracting. It is only relative to the measurements of another observer that a difference appears.

As long as we are in smooth relative motion, we will continue to disagree on measurements of space and time, to argue about the simultaneity of events or the synchronicity of our clocks. But we must agree on the occurrence of events. If a bomb goes off and destroys a building, we will agree the building is rubble. If a child is born, we will agree on his existence. Events happen unambiguously. As for when they happen and where, all we can ever know is when and where they happen relative to whom.

A natural paradox arises (Figure 4.3). If one person travels in a spaceship at near light's speed for a few light years and then returns home to rejoin their twin, both twins believe the other's clock runs slower, so who is actually younger at their reunion? According to special relativity, our motion is totally relative. Each would see the other's time dilate and the other's measure of space contract. But the reunion of the twins is an unambiguous event. They will be able to look each other in the eye, talk, exchange stories. Undeniably, one will be decades younger than the

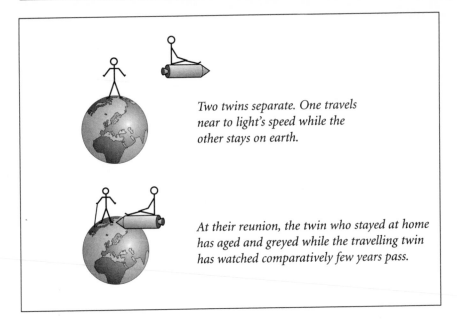

*Two twins separate. One travels near to light's speed while the other stays on earth.*

*At their reunion, the twin who stayed at home has aged and greyed while the travelling twin has watched comparatively few years pass.*

Figure 4.3 *The twin paradox.*

other. One will be grey and might have shrunk a bit and will complain of wrinkles, sagging and other side effects of ageing. The other will have experienced the passage of only a few years. She will look at her aged image in the face of her twin.

The resolution to the twin paradox lies in the limits of inertial motion. In order for the twins to make a comparison of their ages, the one in the rocket would have to stop, turn around and accelerate back up to near light's speed. The principle of relativity does not apply to motion under the action of forces. Firing rockets are not inertial, since they exert a force to change direction and speed, and the equivalence of two observers is broken when one of the twins experiences the forces of the rocket. By carefully comparing the clocks of the twins it can be deduced that the space-travelling twin is younger at their reunion.

Our family is riddled with twins – there was dad's father and his twin sister, also another great aunt and her twin brother, my great uncle. Including cousins and family through marriage, there are at least five sets of twins in the extended family and maybe as many as eight sets, depending on who you include in the count. But the twin paradox has always made me think of the family's highest twin achievement – the identical twins, you and Harriette, my mother and my aunt. The tyranny of the twins would have been weakened by separation. The time dilation is real. If you watched your twin move away in a rocket, you would see your sister ageing slower, operate her ship slower, experience an elongated habitual day. When Harriette stopped in her rocket, turned around and finally made it back to earth, you could look in her eye, your identical twin, and find her decades younger than your own person. To each of you those years passed like any other, nothing seemed odd or was odd, only it was a handful of years to Harriette that passed, not the decades you lived. The two of you wouldn't be able to stand the separation.

### 9 JANUARY 1999

I dreamt I was an astronaut. My dream was replete with references to Samson's curls and my own, modern spacecraft, and toy guns. I can't imagine what Freudian analysis would make of that collage. I don't remember the details, naturally, but I do remember being forced away from boarding my shuttle, soon to be launched into outer space. I searched the blue-black sky for the flicker of my silver spacecraft. What do you think it means?

Technologically speaking, rockets can't yet travel at light's speed. But if they could, I could travel the galactic neighbourhood for a few years. I'd come back and you'd all have aged decades. What a dilemma futuristic astronauts would face. They would have to say goodbye to their parents, family, friends. After a few months in space they would know their parents were too old to still be alive and after years they would know they had outlived their spouses, their children and even their grandchildren.

Rockets can't yet go that fast, but subatomic particles can. There are short-lived particles that always decay after a very well-defined lifetime. With particle accelerators we can speed these subatomic particles to near light's velocity and watch their internal atomic clocks slow, so that the particle's life is extended relative to the laboratory clocks precisely in accord with the predictions of special relativity. Relativity works in the sense that it is predictive and that the predictions are confirmed by experiment.

### 7 FEBRUARY 1999

The mushrooms I got from the store are huge. I slice them like meat and fry them in oil. The kitchen is narrow and filthy. We can't seem to get rid of the dirt. I wipe at it weakly or mop it, but never wash it away well. Warren makes the most progress. It is always black outside. The windows serve as mirrors and I study the curiosity of the rare vision of me cooking under the bright spot of light from the ceiling.

We were in London today. Like some cheap science-fiction stunt, the sky went suddenly black. People scattered as though Godzilla might actually step through the heavy clouds. The sky cracked open, unleashing promised buckets of ice. We thought people were throwing rocks at us when the hail came down. Well it's better than rain, we agreed, and laughed at our own stupidity as we ran in no certain direction.

We're back in Brighton and between the sizzling oil, the glaring light and the changes from age I note in my own face, I marvel at the unstoppable passage of time.

No one really knows why time is distinct and peculiar. Whether we understand it or not, it sweeps us aggressively in its flow. We all move in time, relentlessly forward. We locate ourselves not just in space but also in time. Thousands of years ago England was part of Rome. I can stand here, where the Romans stood in space, but separated by millennia. The Romans and I may have occupied the same location in space, but we have not occupied the same location in spacetime. Time is like a fourth dimension and we often discuss living in a $(3+1)$-dimensional spacetime, three space, one time.

As much as we try to make time the same as space, it still seems different, different enough that we continue to give it its own name. For one, we cannot move freely in time. We cannot, for instance, move backwards. The arrow and direction of time are still mysteries that philosophers attend to more often than physicists. There are some who argue that time does not really fundamentally exist, but this is another issue altogether.

**9 FEBRUARY 1999**
I ran along the pebble beach today. The ocean was flat and bright with a reflection of the sun. I ran and ran and felt the same thoughts circulate through my brain over and over and over again. Home, bath, food, bus. This day I dressed and left for work as I have for the past several days. And so it goes that we start to have a daily life, a routine, and the fever from the move has receded from our minds. I try to make it to work today but only make it to the far end of the street, herded by gates to go the farthest, worst way around. The wind tries to distract me and the rain tries to drive me to the bus shelter. I could chase the bus and search

for coins to cover the fare and navigate for a seat, but instead I'm standing here in everyone's way. Facing against the herd. As though dislodged by the rattling of the weather, a link that's been missing in my research falls into place. It's not the most important thought, not even worth recording. But it is the first sign of the numbness subsiding. I could make it to work, forget these associations, and maybe only the vague feeling of discomfort will hang on me during the day. I push past more persistent commuters and clamber back home. Back inside. On the yellow couch. Ugly stained. Metal bars sticking out, mocking the soft part of my thighs. I sit there on that uncomfortable ugly thing. The windows are coated with dirt from the exhaust of cars and the rain scratches the panes as though trying to get at me. Hours pass. I make little progress.

I try to find a simple expression for my ideas. I figure if there is none, the ideas might be wrong. When I first started to work on topology I wondered about complex properties of spaces and didn't take my own suggestions seriously until I realized the simple way to ask the question: is the universe infinite? Einstein's simplest insights were profound. The simpler the insight, the more profound the conclusion. Before I get to explain topology and manifolds I have to tell you the rest of his vision.

Here are some last critical elements. Einstein also found that inertial mass grows at high speeds. It would take a bigger and bigger force to make any mass move near light's speed. It would take an infinite force to push any mass all the way up to light's speed. We can never catch up to a light beam and witness time stand still. No mass, in fact no information whatsoever, can travel faster than the speed of light.

This is absolutely one of the most profound predictions of special relativity and eventually led Einstein to overturn Newtonian gravity. The fact that nothing can move faster than the speed of light ensures basic rules of cause and effect. I cannot affect something unless I can communicate with that thing. To move a chair I have to walk over to it, touch it, move it. To relay any information to you I have to lift the phone, call you, yap into the receiver. All of these actions move slower than the speed of light. Signals are bounced off satellites, dropped into houses and beamed out of illuminated boxes. We communicate worldwide at stunning speeds but always less than the speed of light. I cannot change the course of events half way around the world without somehow encrypting my intentions on a messenger signal, whether it be as slow as a letter or as fast as e-mail. The finite speed of light defines our past, present and future.

Einstein did not deal with gravitational forces until 1915. When he did finally turn to the unification of relativity and gravity, something truly spectacular would emerge.

The gravitational pull of astronomical objects forces apples to fall to the earth and planets to orbit the sun. Effective as Newtonian gravity is, there is something disturbing about the idea of the sun pulling on the earth from such a great distance or the earth pulling on an apple metres above the ground. This action at a distance is a bit like bending a spoon without touching it. In a natural philosophy devoted to the determinism of cause and effect, this action at a distance should seem alarming, but gravity is too familiar for us to take notice. Except Einstein did.

Newtonian gravity is particularly at odds with his theory of special relativity, which asserts that nothing can travel faster than the speed of light and so nothing can act at a distance. To make gravity and special relativity consistent, Einstein invented the general theory of relativity. He discarded the notion of gravity as a force and replaced it with a theory of curved space. Matter and energy curve space, and our experience of gravity is our fall along these natural curves. The apple is not

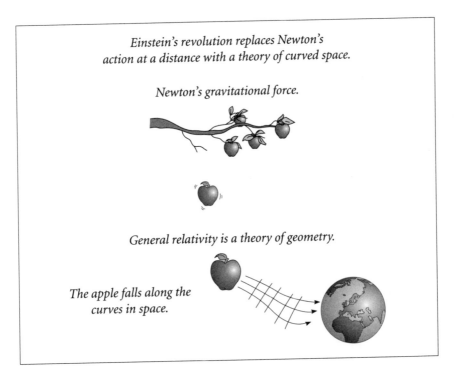

*Einstein's revolution replaces Newton's action at a distance with a theory of curved space.*

*Newton's gravitational force.*

*General relativity is a theory of geometry.*

*The apple falls along the curves in space.*

Figure 4.4 *The geometry of the universe.*

pulled by the invisible hand of the earth; instead it falls along an unseen but nevertheless real curve in space. Action at a distance was overthrown and in its place emerged a theory of geometry: that is, a theory of curved spaces. Spacetime itself became something (Figure 4.4).

If spacetime is real, we can start to ask what it means for it to be not just curved but finite. Cantor gave infinity a precise mathematical meaning so that we could handle it and really examine the operations of infinity with his transfinite arithmetic. Einstein gave space and time mathematical and physical meaning. Once we learned to handle it – and it has taken us the better part of a century to do so – we could start to ask real questions about the shape and extent of the whole universe. Standing on his shoulders, we've been asking those questions ever since Einstein published his theory of general relativity in 1915. The difference now is that we have the technology to glean the answer from satellite observations of the sky. That's what I'm doing on this yellow couch in Brighton, England. I'm not just wondering if spacetime is finite, I'm wondering if we can look out into space and see the whole thing.

# 5

## GENERAL RELATIVITY

**3 APRIL 1999**

Cambridge is heavy and beautiful. There's a big square, a park with long walkways cutting it into diagonals. A swarm of birds scare us. Occulting the clouds they fly and dart, jarring unexpectedly but never breaking formation. Where are they going? Which one is leading? Warren and I are a swarm of two. We sit on the damp grass and go through our options again. I have declined the lectureships. We could make Cambridge work, he says. I think he'll be isolated. It'll be good, he says. I think it'll be hard. He can find work, he says. I try to imagine what he'll do. But how can I not take the job? What are our options? Cambridge is possibly the best place for relativity in the world. You can write a book, he reassures me, my secret step towards independence from academia. I dream of staying still, choosing my home, having a home. Building a life somewhere, not just in my own head. We can spend a few months in California, I reassure him.

So there you have it. No more flow charts, no more pow-wows. We're moving to Cambridge. We keep moving. Moving, moving, moving.

**23 NOVEMBER 1999**

I confess my affections for this royalist island, but I despair the absence of early-American themes in November. I guess they are less inclined to give thanks for our defiance of British rule. All the same, no one is openly moody about the holiday and some friends invite me to London for an impromptu Thanksgiving ceremony. All of the participants are expatriates, but few are from the States. Most of the group share in common their aim to quantize gravity. As with many in our set, there is an innocence in their manner and ideas. A pretty fascinating bunch, I enjoy their company despite my claim to reject academics and embrace anti-intellectualism. Something feels like it's happening in London, as though people are starting to rain on that city with far-reaching and interwoven interests. I wonder if we can maintain our numbers and form a real London group.

I have this hilarious ride home. On the train to Cambridge, I see this guy peering at me from between the seats, bobbling his head around to get a better look at me. From the piece of his face I can make out, I recognize this nutty mathematician who I have serendipitously bumped into every place I've ever been for the past twelve years. I can't believe my eyes. We can't remember all of the times we collided. Was he at Harvard when I was at MIT? We share calamitous stories about that time and laugh with sympathy at each other's plague of embarrassments. Didn't we knock into each other at Princeton next? Then Berkeley? We aren't boasting about our academic trajectories, we're wearing that history like war wounds. Here he is, arms hooked over the seat in front of me with a delirious giddiness, laughing like he could cry on a wobbling old train from London to Cambridge. We part in Cambridge with the full expectation of spotting each other somewhere else on the globe. Nostalgia forges a bond between us and I watch fondly amused when he takes to running home in his dinner suit with a collegiate pack lurching about his back and wrinkling his coat.

I make my own way home, walking not running. I wonder if Warren is up waiting for me, rehearsing, antagonizing our new neighbour with his commitment to musical detail. As it happens, our old neighbour didn't much care for American bluegrass. His revulsion has literally driven him out of his home. He sold, packed and moved.

I walk home and admit to myself unwillingly that the absent-minded scientist can soften your heart. I remember the stories of how Einstein stumbled and struggled to unify gravity and special relativity. Was his hair frayed and his moustache twitching when deep in thought? Did he really forget where he lived or his phone number? Did he really say, 'I never remember anything I can write down'?

Absent minded in some ways, unyielding in other ways. He may have been cavalier with his phone number, but he was relentless in his attention to logic. We've only touched upon Einstein's theory of gravity and there's more we'll need to grasp before we can progress. Starting where we left off: despite the impeccable condition of Newton's theory, Einstein realized that gravity violated one of the sacred tenets of special relativity. As mentioned, the main conflict with special relativity is that Newtonian gravity allows information to be communicated faster than the speed of light. If the sun were to suddenly collapse, according to Newton's theory we would immediately feel the gravitational change. This is no different from expecting an event in Iceland to instantaneously alter the course of events in China. Einstein's theory of special relativity forbids anything, any form of information, to travel faster than the speed of light. Information of any kind must be encoded in the form of energy or mass, after all $E = mc^2$. Information is encrypted on stuff, and stuff travels slower than, or at, the speed of light. The force of gravity should be no exception. Any information about a gravitational change should travel slower than, or at, the speed of light. To ensure this, Einstein had to discover the essence of gravity and I'll try to give you a peek at that essence here.

Probing questions expose the fact that no one had a clear, tangible understanding of gravity before Einstein. Newton wrote down a mathematical expression describing how all mass attracts all other mass, but he still had to confess that in some sense he didn't fundamentally understand what this force was. Before Einstein, we had a set of rules for describing gravitational attraction but no real sense of how those rules were enforced.

Newtonian gravity is fantastically successful. He brought the stars, the sun, the weight of our own bodies into unison by realizing that all had gravitational attraction in common. No matter how obvious hindsight had made Newtonian gravity, Einstein was still needled. He pulled back at the layers of the theory until he realized that Newtonian gravity was a humble façade and that beneath the façade the universe was truly strange, stranger than his wildest imaginings.

There were two important realizations that Einstein came to with deep intuition. First he defined a principle of equivalence and second he made the magnificent suggestion that spacetime was curved.

The principle of equivalence alone is powerful. When we walk around the surface of the earth we feel the effect of gravity resisting our feet; it gives us the impression gravity is pulling us down towards the earth's centre. Einstein realized that when an elevator lifts us up several storeys in a building, we experience the same sensation. The floor of the elevator resists our feet and as we accelerate upwards we feel pulled towards the floor of the cubicle. So he made this leap: acceleration is equivalent to a gravitational force. That's the principle of equivalence. So there he goes acting all simple. Making simple declarations that he then clamps down on with the tenacity of a rabid pit bull. He pursues the consequences to the death, the death of Newtonian gravity.

After settling into this insight, he decided that gravity isn't really causing an acceleration, but that the resistance to gravity causes an acceleration. Leaping from the top of a building, a body is, for the few moments before impact, in complete freefall. The freedom of freefall, the vertigo of freefall. The body experiences no force, a suspension of forces, as though in outer space in zero gravity. We don't experience a force until we hit something along our natural path, a chair, the floor, the crust of the earth.

So what is gravity if it's not a force? It is the shape of space. Gravity is a field and mass is the charge. An electric charge creates an electric field around it proportional in strength to the amount of charge. Mass, like electric charge, creates a field around it in the form of a curved space; a gravitational field as opposed to an electric field. Empty space is flat, but in the presence of mass or energy of any kind, say a star like our sun, space will curve around it and time will warp. We fall to the earth by following a curve in space and the earth orbits the sun along space's bends. It is impossible to visualize a bent three-dimensional space, but we can visualize a bent two-dimensional space to help us along. A bowling ball nestled on a canopy is a popular analogy (Figure 5.1). The mass curves the sheet so that a marble rolling on the curved canopy will fall towards the bowling ball, or it can be spun around the bowling ball into an orbit. Like any analogy it's not perfect, but it does provide a powerful image.

We are intrinsically bound to this space and this time. We cannot jump off it or live outside it. We are meaningless without it. This is our universe – the vast extent of our curved spacetime. People always ask: what's outside the universe? The answer is nothing. It lives nowhere. There is no meaning to the questions where or when if there is no space or time. The tendency is to imagine a curved space as bent into

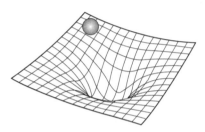

Figure 5.1 *The moon falls along a natural curve made in the shape of space by the mass of the earth.*

something, embedded in another space. Whether or not that's the case, it is totally unnecessary. It is completely possible that that is all there is: (3 + 1)-dimensional curved spacetime.

Mass curves space and we all fall freely, without force, inertially, along those natural curves in space. Any body will take the path of least resistance along a curved space, rolling along as though there is no force at all. Planets orbit the sun by following an elliptical path defined by the natural curves. Apples fall to the earth by breaking loose from the pull of the tree and following the path of least resistance along the invisibly curved space until the surface of the planet interrupts its fall and forces it to stay still. Between the tree and the planet, the apple is inertial, free of forces. We feel gravity as a force only when we try to resist it, hang from a rope or hit the floor.

This is a more general theory of relativity than the special theory and so sensibly has grown into the name general relativity. Gravity and relativity are made compatible in general relativity. If the sun were moved or were to disappear, space would ripple and move as a wave in the shape of space, a gravitational wave, propagating the information about the relocation of the sun at the speed of light. Gravitational waves are like the gravitational analogue of light waves and share their limiting speed.

Space and time are still relative. Different observers will measure different lapses in time and different distances between events. However, general relativity incorporates the warpage of the fabric of spacetime in the relationship between different measures of space and time. Unlike special relativity, which we now see as a theory of flat space, all observers are not equivalent on a curved space. Spacetime is warped, so observers in different locations know they are in different degrees of warpage. Clocks will run slower for an observer in a highly curved region of space relative to an observer in a flatter region of space. Observers closer to the sun will use clocks that run slower than those further away.

And here, I'm home. I struggle with my key. The lock is too close to the edge of the door, forcing me to scrape my knuckles on the raw bricks. As the door opens from the inside, I fold up my thoughts on general relativity and Einstein and come in from the cold.


**1 DECEMBER 1999**

I wonder if Einstein would do a victory dance. We used to do victory dances in graduate school if we managed to understand something. We'd prance around the room with our arms and elbows vigorously extended, hyperextended if you had the flexibility, but otherwise rigid as boards. I wanted to implement the chest-to-chest slamming celebration preferred by professional basketball players, but the boys were too intimidated.

Einstein did it, victory dance or no victory dance. Newton wrote down equations that we now think are only approximately correct. Newton's laws describe the rules of attraction between masses when space is weakly curved. Einstein had no equation for general relativity yet. He didn't know what equations to write down to describe strongly curved space. Armed with the crudest mathematical instruments, he pierced the surface and saw through to the core. First he identified the object of pursuit, geometry. Then he realized he was completely unequipped to handle a battle with such complex geometry.

He had scoffed at the mathematicians for formalizing special relativity. He thought their elaborate methods obscured the simplicity of the subject. With general relativity he realized he was seriously disadvantaged. He did not have the mathematical tools he needed to precisely describe how mass and energy would curve space.

He grappled alone for a time before turning to his mathematician friends. He first learned that an abstract and difficult mathematics had been invented by Georg Friedrich Bernhard Riemann (1826–1866) in the nineteenth century. Riemannian geometry handles curved spaces in any dimensions. As much as Einstein would have liked to live without it, he was forced to adopt this mathematics of curved geometry. Riemannian geometry is demanding and beautiful. Einstein struggled to master these tools and cast his theory into a Riemannian formulation.

Einstein had created an unwieldy monster that in a way he couldn't tame. He conjured up a theory reliant on mathematics in a curved spacetime that still demands years of its students' attention. Though he managed to use these tools, compared to his mathematician friends he used them clumsily. Isn't that great? I love that. His fragility. His defiant brilliance in the face of his own limitations. He still did it. He came up with relativity and not the mathematicians; although David Hilbert (1862–1943), Marcel Grossmann (1878–1936) and Hermann Minkowski (1864–1909) made enormous contributions, it was still down to him. He ploughed right past his inadequacy. Maybe this is what he meant when he said 'Imagination is more important than knowledge.' Like a bad plumber he hacked and hammered and slapped together a mathematical model for curved space, correcting error after error in his own formulation. Sloshing between despair, doubt and conviction. When he finally pulled something together, something that worked, he was overcome with elation for days. He had trudged through the darkness of his own confusion and found what he set out to discover; a theory of gravity based on curved spacetime and faithful to his principle of relativity. It's like Michelangelo revealing the sculpture he believed hidden within each stone.

What could this intuition be? Special relativity is phenomenally simple mathematically and relies on the simplest, clearest physical arguments. As though the rest of us live in this fog that he could just see through. He followed his intuition like a beacon, distrusting his calculations but not faltering in his faith. Where does this kind of knowledge come from? Is it there in his mind? In my mind? Yours? Waiting to be mined? Are all of nature's greatest secrets encrypted in our own selves? I hope so. I think so.

So there, am I doing it? Hero worship. If I ever had a hero, it would have to be Einstein. I know how cute he looks with his electrified hair and taunting smile and warm eyes. I also know he may not have been so cute in real life. Maybe he was a no-good man or a bully. Who really knows? But what's the point of thinking like that? We all need heroes and he's mine.

I can't find any evidence of real insanity in his biographies. Not that I'd want to. Not everyone can be a great source for my morbid curiosity. I have heard that one of his children was mentally impaired and that he had little paternal contact with him. He does not to my knowledge have a genetic legacy, in the sense that none of his children followed him into theoretical physics. His genius in theoretical physics stopped there, with him, at least biologically. But intellectually he literally opened the universe to every scientist after him. People more agile in mathematics began to make progress where he couldn't. What a legacy he left us, if not his children.

## 3 DECEMBER 1999

Pedro Ferreira and I worked together in Berkeley, crunched in a box of an office so small our chairs would catch if we tried to make use of their wheels. To talk we only had to lean back and we were face to face. This is where our friendship began. The paint peeling from the ceiling. The bolts in the wall. I put those bolts in the wall, says Pedro, admiring the shelves he had installed and burdened with books. We were inseparable when we worked at Berkeley, but only between morning and evening, and our personal lives were led completely separately although I liked to give him all the details of my own.

Both of us were new to California and we'd compete over who could resist assimilation the longest, the most ostentatiously. Like most of the researchers in our department, we are uprooted and moved every couple of years. It's a very peculiar aspect of the scientific culture, this reshuffling of postdoctoral fellows. We spent a good fraction of our time worrying about the strains on our transported families and companions. We've rearranged again since California. Pedro's now in Switzerland at CERN (Conseil Européen pour la Recherche Nucléaire, now the European Laboratory for Particle Physics) while I'm in England. We talk on the phone ritualistically. Ideally we talk three times a day: once in the morning, once or maybe twice during the day, and if we can sneak it in without offending our partners, we have a call to close the evening. Pedro and I are rarely physically demonstrative of our affections. Our salutations are only awkwardly accompanied by a hug. The one time I've held his hand he was surprised at how bony my paw felt in his.

Pedro and I have taken very different academic paths. From the outside they might not seem so dissimilar. We both work in the arena of the large-scale properties of the cosmos. These are things Pedro and I take for granted: space and time can be curved and stretched, can be born and can die, begin and end. Spacetime is not matter or energy but a field responding to both.

Newton's laws assert that in the absence of forces an object will move along a straight line. In Einstein's theory of curved space, objects and light beams follow the straightest path they can find. Light and matter are lazy and take the path of least resistance, and on a curved space, the path of least resistance will not be strictly straight.

Spacetime will warp in response to the distribution of matter and energy. Our experience of gravity is a free-fall along the most natural curves in space, which keeps the moon winding around the earth and the earth around the sun and the sun around the centre of the galaxy and the galaxy on its as yet unknown path.

The geometry of spacetime will depend on the nature of the matter and energy shaping its curves and mutations. A round lump of mass carves a different curve into spacetime than a smooth distribution of matter. A collapsed lump of matter leads ultimately to the unheralded prediction of black holes. A smooth distribution of matter drives an expansion of space and leads us to a realization that the universe had a beginning. Einstein did not foresee or welcome these predictions of his theory. Black holes and the big bang are the bizarre yet unavoidable fledglings that others discovered in the wake of his revolution.

Black holes first, which brings us to Schwarzschild. During volunteer military service, a young German mathematician, Karl Schwarzschild (1873–1916), found a solution to Einstein's theory in the presence of a round concentration of matter, like a star. The solution describes the geometry of space around the ball of mass and the curves in this geometry were also found. The curves define the paths of inertial motion and so describe the orbits of bodies around the central source. Kepler's laws would have all of the planets orbit on perfectly closed ellipses. Peculiar observations that predate Einstein found an anomalous shift in the orbit of the planet Mercury. The ellipse that Mercury followed appeared to lag, never quite returning to its starting point. This precession of the orbit was only very slightly off from the Newtonian prediction. Still, Schwarzschild's solution predicted that planets would not orbit on closed ellipses, but rather that precession of the orbit is to be expected. The relativistic prediction that correctly gives Mercury's orbit was one of the first observational successes of Einstein's theory.

In addition to correctly predicting the anomalous precession of the orbit of Mercury, general relativity predicts that the path light follows would be bent very slightly as it passed by the sun. This is nearly impossible to observe given that the sun blinds us with light. However, the moon just happens to subtend precisely the right size in the sky that it can completely occult the sun in a total eclipse. The darkness of a total eclipse provides the necessary backdrop for observers to measure the tiny deflection of the image of a distant star. A few years after Einstein's prediction of the bending of light, the effect was confirmed observationally.

There was a partial eclipse visible from Europe on 11 August 1999, just before we left Brighton. The moon cast a shadow over the earth and the day turned not quite dark. I would have to say, the air seemed to change colour – bluer or browner, it's hard to identify. The tiny spaces between the leaves of the trees acted like pinhole cameras, projecting thousands of swaying images of the crescent sun on to the pavement. We wandered around Brighton in a stunned silent crowd, looking at the ground decorated by the sky. We actually miss Brighton. Hard to believe since it was such a hard year. We miss the beach. We wanted to take a dictaphone around the pier to catch the haunting funfair music that makes me think of scary clowns. We'd hang over the railing of the pier that extends out into the ocean and listen to the clamour of pebbles, the competition of a million voices.

When I turned down a job recently, citing personal reasons, they told me to lose the boyfriend. I actually thought it was pretty funny. As though being located on some other point of the earth was going to get me closer to the action. What's a few thousand miles, give or take, compared to thirty billion light years? Brighton is as close to the big bang as any place.

The big bang happened everywhere, creating a curved space we call home. The simplest kinds of curved space are conveniently also relevant to cosmology. The shape of spacetime – that is, the shape of an entire universe – is thought to have roughly constant curvature everywhere. I say the shape of 'a universe' because we don't know if ours is the only one out there, and already I've misled you by using words like 'out there', which imply a direction and a location. But if they are truly separate spaces then by definition they are not linked to us by space and so are neither 'out' nor 'there' relative to us. They are completely distinct. We can never communicate with separate spacetimes. All energy and matter move along space and we cannot communicate unless space is contiguous and provides a bridge for our matter or energy signals. If any such bridge exists between warped spaces then they are to all intents and purposes one universe. So I use the word 'universe' to mean the entire connected spacetime.

The simplest solutions for the large-scale landscape of a cosmos are three spaces of constant curvature: the universe may be everywhere flat, everywhere positively curved or everywhere negatively curved (Figure 5.2). To understand the shapes of these spaces we have to bend Euclid's rules. Euclid of Alexandria (about 325 BC–about 265 BC) formulated postulates of triangles, lines and polygons on a flat surface. Over 1,500 years later, Riemann formulated non-Euclidean geometry, which describes

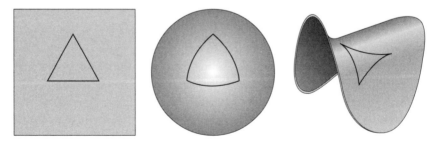

Figure 5.2 *Triangles drawn on a flat surface have straight edges and inner angles that sum to 180 degrees. Triangles drawn on a spherical surface bow open, and on a saddle with negative curvature they narrow.*

the nature of triangles, lines and polygons on curved surfaces. Triangles taper on saddles and open on spheres. They look like Warren's hands and feet. Triangular hands are ambiguous. They appear to open when offered to you and close into a fist when taken away. I can't be sure if they are opening or closing; if they are keeping something out or keeping something in.

Flat space in its simplest incarnation is infinite, like an infinite sheet of paper. In flat space, light travels in straight lines. We could measure circles, triangles and the paths of light beams to deduce that space looked flat in our local neighbourhood.

The everywhere positively curved space is the simple sphere, like the earth. On a sphere, light will travel along the natural circles – a ray will not spontaneously deflect into a crooked path but takes the straightest path allowed. A light beam confined to a sphere could be directed to a near neighbour, in which case it will cross a short arc before being intercepted by that neighbour, or it could be directed away from the neighbour, in which case it will travel along the much longer arc before being intercepted (Figure 5.3).

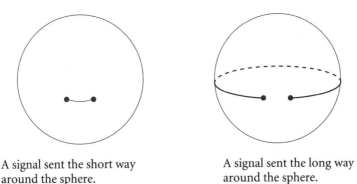

A signal sent the short way
around the sphere.

A signal sent the long way
around the sphere.

Figure 5.3 *On a curved surface, light takes a curved path.*

Triangles bow open when drawn along a sphere (Figure 5.2). The amount of distortion will depend on how big the triangle is drawn relative to the curvature of the surface. A very tiny triangle will appear flat, while a very large triangle will not, clearly showing the positive curvature.

It has been intuitively understood for ages that geometry is different on a curved surface than on a flat surface. There is a remarkable painting by Jan van Eyck from 1434 called *The Arnolfini Wedding*. The painting is often referenced for its immaculate depiction of non-Euclidean geometry. On the far wall behind the newly weds hangs a round convex mirror that reflects the entire scene. Among the many notable features of this painting is the rounded distortion of the images on the mirror's spherical surface.

The negatively curved surface called the hyperbolic plane cannot be drawn properly on a flat sheet of paper. We can draw one local neighbourhood of a hyperbolic plane to illustrate what negative curvature tries to do to the shape: it tries to create a saddle at every point. We cannot draw a saddle at *every* point on our flat sheet of paper, but we can deduce that such a space could exist if it were not forced to embed itself here. The hyperbolic plane is infinite – in other words, it is unconnected and can stretch forever, in contrast to the sphere, which is compact and finite.

On a negatively curved space, light veers along the saddled orbits. Two light rays emitted close together will quickly diverge along these curves. As a consequence, the interior angles of a triangle narrow, so triangular corners appear pinched on this negatively curved surface. The infinite hyperbolic plane is larger than the infinite flat plane. Strange but true. We'd have to shrink and distort the images on the hyperbolic plane to fit them down on to the flat plane. Although no simple mirror can imitate negative curvature, any plain shaving mirror or warped reflection will bend the rules of Euclid's geometry.

These three solutions are central to cosmology. The observable universe appears to take the form of one of these solutions, at least locally in the region we observe. Which one of the three is still open to debate.

Despite its great success, even Einstein's theory is incomplete. General relativity does not fully predict the geometry of space. It does offer a theory for local curves, but it does not determine the global shape and connectedness of space. Relativity cannot distinguish between a universe that continues forever and one that wraps back on to itself, is finite and edgeless. The global shape and connectedness of space is the domain of topology, a branch of mathematics that has seen profound advances this century. Even though we still don't understand the deepest connection between matter, energy and gravity, we know much about what it means to live in a spacetime. From this perspective, as observers and performers of thought experiments, we can chart out the space in which we live, just as we have charted the oceans and continents of our planet. This might seem limited, imposing our human perception to try to deduce the grandest cosmic code. But we are the product of this universe and I think it can be argued that the entire cosmic code is imprinted in us. Just as our genes carry the memory of our biological ancestors, our logic carries the memory of our cosmological ancestry. We are not just imposing human-centric notions on a cosmos independent of us. We are progeny of this cosmos and our ability to understand it is an inheritance.

# 6

## QUANTUM CHANCE AND CHOICE

**11 DECEMBER 1999**

Sleep releases the latch I keep fastened by day and my mind floods with a sea of images I can only fend off when awake. Am I keeping things out or keeping them in?

One dream keeps invading me. I'm trying to get through my day and snippets fly at me from nowhere. A door. A floor raising instead of an elevator. An old dorm room with a strange key. Collections of keys, but that's waking life. In reality I have collections of keys to so many doors, apartments, labs, offices. Front doors, side doors, back doors. Double locks, British keys, big keys, little keys. I forgo my deposits. I'd rather have the keys. They collect in heavy lumps in random bags and pockets, accumulating with each move and collectively forming a jigsaw puzzle which, when put together, makes a map of my random path through the world.

This dream came in pieces, segments, forming over the last several weeks. Strange buildings and elevators dropping, giving me Einstein's freefall so convincingly that I press my head to the glass to settle my stomach which isn't used to zero gravity and resign myself to the fate of my fall but I wake up and quickly forget only to remember in bright flashes while I'm trying to talk so my speech slows and my eyes strike a diagonal while I wait for the image to fade and try to recover my place in my sentence like finding my lost place in a book.

But I'm okay by day and it is day now. On this particular day, it is a typical day and I'm reading a preprint written by someone who works at Fermilab near Chicago. Before we all started distributing our papers on the internet, each department would run off xeroxed copies of their preprints, prior to publication. Fermilab's covers were yellow and we used to call the massive number of Fermilab preprints covering our journal racks the yellow plague.

Fermilab produced a generation of particle-astrophysicists, people who study the overlap of physics on the largest scales with physics on the smallest scales. That's where my graduate adviser Katie Freese received her PhD under David Schramm's supervision. Katie is a striking woman with an equally notable personality. We were not always easy collaborators, and made for an amusing, if turbulent, adviser/student pair. Her adviser David Schramm was a force in modern cosmology who refused to go to a conference unless he could fly there in his own plane. My academic grandfather shockingly and tragically died in 1997 when the twin-engine plane he piloted crashed on his way to the Aspen Center for Physics. He will be remembered as one of the monumental personalities in particle-astrophysics.

Even though cosmology concerns physics on the largest scales, we can't ignore quantum theory, the theory of the smallest constituents of matter. There was another coup brewing in theoretical physics around the same time as Einstein devised the theory of relativity. Relativity revised our concept of the universe on the largest imaginable scales, the very essence of space and time. Meanwhile, back at the ranch, the universe on the smallest imaginable scales, the microscopic arena of matter, was proving to defy common sense. The name for the modern view of microscopic physics, quantum theory, comes from the idea that all matter and all energy come in small discrete units, quanta. We might look at a bath of water and see a continuous flow, but on the atomic level water is a collection of individual atoms and the atoms themselves are constructed from individual quantum units like electrons, protons and neutrons. On the microscopic level, no physical thing is continuous, all of the world is quantized.

Quantum mechanics is truly weird. There is an element of truth to the rumour that no one really understands quantum theory. Still, we can use the theory. Even though most of us don't really understand the mechanisms behind our televisions, phones or computers, we still watch, talk and type. Many of us brazenly use quantum theory with the same confidence despite the confounding way that the theory insists life is not as common experience would dictate. Life is so weird according to quantum theory that not only is it hard to understand, at some level it seems impossible to understand.

Relativity may be difficult to come to terms with at first and it may violate the intuition of our everyday experience, but it is not impossible to understand. Quantum mechanics presents logically opposing situations, mutually exclusive states, and then insists that they coexist. So in some sense our minds cannot find the truth there. Despite the seeming impossibility of the assertions, the theory nonetheless makes extremely accurate predictions and hasn't failed any tests.

Quantum theory says we are here and not here. We are made of particles, no waves. Neither, both.

In the classical picture, light is a wave, a smooth and continuous flow of electromagnetic energy. When viewed as a wave, the higher the energy of the electromagnetic radiation, the shorter the wavelength. The lower the energy, the longer the wavelength. And though our bathroom taps denote cold with blue and hot with red, it really is the other way around. The different wavelengths of light appear to the human eye as different colours. The hotter the light, the higher the energy, the shorter the wavelength, which pushes the light to the bluer end of the spectrum. The cooler the light, the longer the wavelength and the more red the light appears, at least if it were in an optical bandwidth that we could see. Our eyes evolved to receive and decode wavelengths of light in a bandwidth emitted most powerfully by the sun. Below the visible spectrum is infrared radiation, which we can see with special glasses to give us night vision. The heat from human bodies radiates in the infrared and, though we can't see our sleeping companions glowing in the dark, we can feel the energy of that heat. There is electromagnetic radiation of much longer wavelengths, lower energies that we simply can't see but can detect with tremendous telescopes, like the spectacular arrays of radio telescopes in New Mexico. Working in concert, these huge dishes rotate in unison to act as giant eyes. The visible spectrum then runs from red through yellow to blue, and beyond to the ultraviolet that we can't see but which scorches our skin, and beyond that to X-rays. We build instruments like X-ray machines to translate the information from this high-energy, short-wavelength light into false colour images that we can see with our human eyes. I tend to use the words 'light' and 'radiation' interchangeably to describe the entire range of electromagnetic energy. A light wave can be described equally well in terms of the frequency of oscillation of the wave, which is inversely proportional to the wavelength. Long-wavelength light has a low frequency, while short-wavelength light has a high frequency (Figure 6.1).

Maxwell's theory of electromagnetism seems to predict that an infinite amount of energy could come from a radiating body like the sun and once again common sense rejects infinity. In short, precise calculations

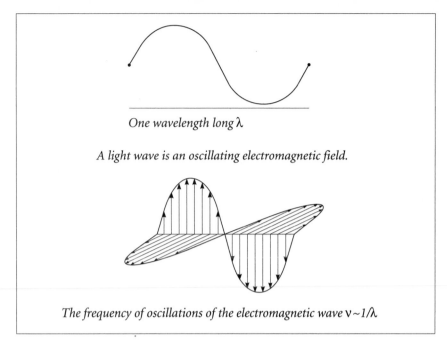

*One wavelength long λ*

*A light wave is an oscillating electromagnetic field.*

*The frequency of oscillations of the electromagnetic wave* $v \sim 1/\lambda$

Figure 6.1 *The wavelength of an electromagnetic light wave and the frequency of the wave.*

using Maxwell's theory of electromagnetism predict that a hot body will emit electromagnetic radiation of all accessible wavelengths and that each wave will carry an equal amount of energy. Since there are infinitely small wavelengths of light, each carrying equal energy, the total energy pumped out by the whole body must be infinite according to these calculations. But of course the sun does not produce an infinite amount of energy, but rather a finite amount, and the emission of light is peaked around visible wavelengths. The infinity predicted in classical electromagnetism was bad enough to result in the demise of our classical picture and usher in the reign of a theory that we don't quite grasp, but that works without giving us unruly infinities. In 1900 a German scientist, Max Karl Ernst Ludwig Planck (1858–1947), modified the picture by suggesting that the light wave does not come in continuous energy denominations. In other words, light can carry only a discrete, quantized unit of energy.

The unit of energy carried by a given wave is related to the frequency of the wave. The higher the frequency, the higher the energy. Before Planck, every wave was expected to contribute the same amount of energy and there was no obstacle to them doing so. After Planck, there was an obstacle. If a given wave carried a unit of energy that exceeded the set energy, it simply wouldn't contribute, cutting out the infinite accumulation of energy emitted. The spectrum that Planck predicted exactly matches the observations of the energy emitted by a hot body. This first step in the development of quantum mechanics earned Planck the Nobel prize for physics in 1918.

Einstein goes one step further and says that not only the energy of the light, but also the light beam itself is quantized. Light isn't a classical wave but is really a collection of individual bundles of energy, quanta, called photons. Like the bath of water, the individual quanta are extremely tiny and the light beam appears to be a continuous swell of electromagnetic waves. The energy is quantized because the electromagnetic bundles are fundamentally quantized.

You can see the particulate nature of light in action. Einstein received the Nobel prize in 1921 not for special relativity or general relativity, but for his understanding of a quantum phenomenon, the photoelectric effect. The photoelectric effect refers to an experiment in which light is shone onto a metal plate to dislodge surface electrons. If you shine a wave on a metal plate, you might expect that as you increase the intensity of the light, more energetic electrons will be freed. This accumulative effect is what you would expect if light behaved as a wave, but that's not what happens. The scattered electrons are more energetic only if the colour of the light – that is, the frequency – is increased. This observation is consistent with the particulate nature of light, the quantization of the lightbeam into photons. Shining photons on a metal plate is like throwing balls at a bull's-eye in the funfares of the Brighton pier. The effect is not cumulative, it does not depend on how many balls are thrown at the target, but instead it takes the right energy of one given ball to hit the bull's-eye and release the funfare prize.

All in all, this isn't that terrifying a prospect. So matter and energy are quantized, big deal – like a Pointillist painting that seems a smooth rendition from afar, but in which up close both matter and light resolve into an array of individuated dots. Only it gets so much more peculiar. Sometimes light *is* a wave. Sometimes light is a particle. These are two mutually exclusive propositions. Either light is fundamentally a wave or it is fundamentally a particle. Quantum theory really forces upon us this peculiar state of being: light is a wave *and* a particle. This wave–particle duality is the first blow to our grasp on the idea of a reality that is independent of the observer.

In a wholesome reality independent of the observer, particles are particles and waves are waves. These should be defining, unambiguous and inseparable properties of a thing and yet they are not. Nature is dual. Waves are particles and particles are waves and the form embodied seems to depend on how the thing is observed. The great unanswered mystery of quantum mechanics circulates around the role of the observer in collapsing this ambiguity. When we observe light and matter, depending on the experiment, they will act *either* as a wave *or* as a particle and the specific form assumed seems to depend on how light and matter are measured. Reality, at least in some aspects, seems to depend on the observer.

### 13 DECEMBER 1999

Over a year from this date I'll be on the Melvyn Bragg radio show *In Our Time* along with Lee Smolin and John Gribbin. Of course, I'm not psychic, I'm just editing. Bragg will turn to me and on live radio, with an audience of a million and a half, ask me to explain the importance of the double-slit experiment to our listeners. It's been ten years since I thought about the double-slit experiment. We had a good laugh over that.

In the experiment, light is shone through two slits on to a photographic plate. If light is made of individual corpuscles, then when the photons hit the barrier with two open slits the only image on the photographic plate will be two thin slits from the impact of the individual particles. If instead light acts as a wave, then when passing through the two slits the wave fans out past the barrier and interferes with itself, like two water waves merging on a pond. The net effect is an interference pattern on the photographic plate. The double-slit experiment when carried out shows that light is in fact behaving like a wave, whereas the photoelectric experiment shows that light is in fact behaving like a particle. The nature of matter is dual and seemingly impossible.

In 1923 a French prince, Prince Louis Victor Pierre Raymond duc de Broglie (1892–1987), combined Planck's quantized energy and Einstein's quantized photons with the special relativistic result that $E = mc^2$ to argue that all matter, though seeming particulate in nature, must also have a wave-like character. The wave is so tiny that we can't resolve this aspect of matter unless we perform high-precision experiments, so we carry on in life as though matter were solid, not oscillating. Still, at the core, the wave–particle duality is pervasive. Matter and energy are *both* particles and waves, and that is not something I believe anyone truly understands.

Matter and energy can simultaneously be here and there, be and not be.

The wave nature of the electron is a probability wave. Quantum mechanics requires the profound state of not-being: the electron does not exist at a precise location in space with a precise energy. Rather there is some probability for the electron to be here or there but no definite meaning to where it is.

The probability wave describes the state of the electron (position, velocity, etc.), but an uncertainty principle due to the German physicist Werner Karl Heisenberg (1901–1976) says we cannot ever precisely pin down the position of the electron without losing other information, in particular without losing knowledge of how fast it's moving. The uncertainty principle says that there is an uncertainty in the location of a particle related to how precisely we have measured its velocity and vice versa. If we try to determine the locale of the electron, we lose information about its velocity. There is a tendency to imagine that the electron has a precise location and precise velocity, but that when we try to measure the whereabouts of the particle we have to disturb it in some way, knock it about so that we lose information about the speed. The quantum mechanical interpretation is more dramatic than that. The electron does not have a precise location and a precise speed until we, the observer, measure it. This phenomenon, known as the collapse of the wave function, has ensnared physicists and philosophers since its discovery. The full import of this conclusion always sends me careening.

This interference between the observer and observed is truly profound. We are part of the system: as austere and distant and objective as we try to be in our scientific investigations, there is a theoretical limit to how precisely we can remove ourselves from our object of investigation. The questions we ask in some part determine the answers. It brings to mind the social constructivists. I don't think I can do justice to a definition of social constructivism, but in brief it is the notion that each of us brings our social context to the world we study, which seems a fair enough assertion, but extremists lurk everywhere and the extremists might say there is no objective reality, just the social context we bring to the world.

I'm not totally unsympathetic to the social constructivist view, but to be extreme – although there are few social constructivists who would take such a position – if there's no objective reality, how can there be a social context or a society of people? How can there be people at all for that matter? This leads us to the dangerous territory of solipsism, which asserts that you are all figments of my imagination or maybe it's me who is a figment of yours.

Has quantum mechanics abandoned us to a terrifying abyss where observers create reality? If so, why are we so limited in the reality we can create? I might be able to make an electron appear in a given location, but I can't easily produce an elephant on my city block. Why are experiments still reproducible? Why are there still rules, like the Schrödinger equation? To a person who believes that nature is a purely human construction (as opposed to the more moderate view that the scientific practice is intrinsically tainted by culture and the human context), the role of the observer in quantum mechanics may come as a source of vindication. I don't think it is. I think the distinction between observer and observed is a profound and poorly understood issue. I don't know the answers, but it does give us some divine questions to ask.

### 17 DECEMBER 1999

A translucent-skinned theologian with ambitious corpuscles staining his face sits across from me at a college dinner. An elderly woman encourages the burning red embers in his cheeks by demanding to know what he thinks of Unitarians since she's a Unitarian, cooing like a shrill bird and waving her hands as if to swish away the giddy thrill of naughtiness associated with her denomination. Recovering from her joyous shame, the radical Unitarian asks me politely, 'And you?' What reply is there to make except 'I'm not a Unitarian. I'm a physicist. And my boyfriend's a hypochondriac.'

It's not that I don't belong here, I mean of course I don't, but I never mind that. I like the ridiculous cultural differences. The mundane, elementary things are given a twinge of the absurd, which takes the edge off daily routine. I don't even mind when they make fun of my American accent, which I mutate for emphasis, drawing on all the influences of my migratory pattern. I like having a stride that crosses class and education, but sometimes I can't keep up with myself or smooth the transitions.

They talk and dither, but the sound fades for me only because I'm in a peculiar mood and I feel protected by thoughts of peculiar things. And I have to finish my story about uncertainty, a major theme in my life at the moment. I find no comfort in the premise that uncertainty may be fundamental.

Heisenberg found a mathematical expression relating the precision with which we could simultaneously measure the position and the velocity of a particle. The expression states that the uncertainty in the location and the speed are inversely proportional, so if we measure the location very precisely, we will have a large uncertainty in the velocity and vice versa.

There is another incarnation of the uncertainty principle, which states that there is an uncertainty in the energy and the time of an event. If, for instance, the energy of a particle is very precisely measured then the time it took to perform the measurement will be correspondingly long. The deeper implication is that a particle does not have an exact energy at an exact moment. This has shattering implications for some tenets of classical physics that previously seemed sacred. For instance, conservation of energy is one of the most trusted rules of classical physics. The conservation law states that we can always – at least in principle – account for all of the energy in a system. Energy never disappears into nothingness or appears out of nowhere. However, the uncertainty principle allows violations of the conservation of energy if the duration of time is short enough. The implication of this is huge. Particles and their antiparticles can bust into existence and drop back out of existence without violating the quantum rules. Something seems to spring from nothing, but only for a very short time. Not only can this happen, but in principle it happens continually, creating a microscopic foam. Energy will appear to be conserved on familiar macroscopic scales, since on macroscopic scales we are sensitive to a kind of average over this microscopic sea, and on average energy will appear conserved. But when probed on microscopic scales, the world is awash with a continual activity of creation and annihilation.

The uncertainty principle pervades all of quantum mechanics and can be used as a rough guide for the rules of this very peculiar game. Among the more surreal consequences of uncertainty is the phenomenon known as quantum tunnelling. The uncertainty in the position of a particle means that there is a small but non-zero chance that a particle on one side of a seemingly impenetrable barrier is actually located on the other side. Quantum uncertainty allows us to tunnel through walls, disappear and reappear with some finite if protectively small probability.

Our modern picture of the atom is also rewritten by uncertainty. Before Heisenberg's 1927 formulation of his principle, there was an historical planetary model of the atom where negatively charged electrons orbit a central nucleus (Figure 6.2). The planetary sketch of atomic structure was proposed by the great Danish Nobel Laureate Niels Henrik David Bohr (1885–1962) in 1913. He was inspired by Rutherford's

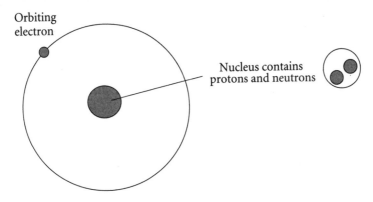

Figure 6.2 *The Bohr model of the atom.*

theory of an atomic nucleus made up of positively charged protons bound together with neutral subatomic particles, neutrons. In the planetary model, different elements such as hydrogen, helium and iron are like atomic versions of different solar systems, with different numbers of electrons orbiting different nuclear centres. Crudely speaking, the electrons live outside of the nucleus and the number of subatomic particles leads to the variety of elements in the periodic table. The planetary model is not very accurate in the end and quantum mechanics gives a very different picture of the motions and behaviours of the electrons and the nuclear particles.

Quantum mechanics says that the electron does not occupy a fixed orbit around the nucleus. Instead there is some probability of the electron being in a given location. This orbital, as it's called, defines a kind of cloud around the nucleus where the electron is most likely to be discovered. The electron isn't a particle whizzing around in this cloud. In a strange sense the precise location of the electron is undefined until we try to measure it. Electrons in an atom can spontaneously appear and disappear on different sides of the nucleus without ever travelling in between. An experimenter would sometimes find the electron on one side of the nucleus and sometimes find the electron on the other side of the nucleus but never in between. So how did the electron get from one side to the other without moving through space to get there? Like a magician's knife that passes through his lovely assistant's body without doing damage to her flesh. Of course, the magician is tricking us, but the electron actually seems to be doing what we would hitherto have thought impossible. The electron tunnels through solid barriers, spontaneously appearing on the other side.

Where does this leave our urge to predict? A theory that makes no predictions is a lousy theory. Quantum mechanics still allows us to make predictions. We can calculate phenomena on the quantum level with unprecedented precision and verify those extremely specific and detailed calculations experimentally. That's the real test of the theory. Even if I didn't believe the telephone mechanism was physically possible, once I picked up a phone, dialled a number and found grandma Eve on the other line, the demonstration would be convincing. She could tell me about how she delivers meals on wheels to the elderly, that she's quit smoking and at ninety still pumps her own gas. Even if we don't believe quantum mechanics is possible, we can smash atoms, build lasers, count photons. It works.

We can predict the probability of getting a certain result from an experiment and with a vast number of experiments we can build a composite picture that confirms our determination of the probabilistic predictions of quantum mechanics. So quantum mechanics still functions, even if it does so in a way that rocks our instincts.

We never perceive quantum peculiarities in our ordinary life because the blurriness of the quantum world is so tiny that we cannot resolve these effects easily. We don't spontaneously pass through our chairs to find ourselves on the floor. Refrigerators don't spontaneously come into existence and then disappear. The world appears solid, knowable, deterministic.


## 19 DECEMBER 1999

I watch a man choose produce from the farmer's stand, the dimpled orange between his ashen fingers. I'm watching him choose. Is he exercising free will, his freedom of choice? He catches me watching and offers a warm spark of white off his watery eyes and a smile he's worn for maybe seventy years. Is it really possible that his choice of the orange is an inevitable consequence of every event since the beginning of time? Is this sentence the consequence of determinism? I choose to pick up the red apple. To put it down. Quantum mechanics says no, the entire universe is not just an intricate pinball machine set into motion with predetermined, inevitable outcomes. Quantum mechanics says there are only probable outcomes. But does this really give me volition? I don't know of any easy way to mesh ambiguity and uncertainty into a profound definition of self-determination. If this life were to repeat over and over again, there would be some finite probability of my making a different set of choices, not out of wisdom or free will, but from sheer chance. Chance is not choice. Which am I going to do now? Pick up the apple? Put it down? I *feel* as though I have choice. If free will is an illusion, it is a convincing illusion.

Even the argument that we need to believe in free will in order to protect society implies that we could change people's minds by believing differently, which implies a choice, which implies a will. If someone collapses on the couch in apathy because they are convinced by some debate that in fact they have no will, then this torpor is also a consequence of either determinism or chance, if not a determined consequence of every preceding event then a random chance fluctuation along a nearly determined trajectory.


## 23 DECEMBER 1999

You can see why the enigma of quantum mechanics tempts abuse. People try to treat quantum mechanics as synonymous with supernatural. But just because we don't understand it doesn't mean all sanity has abandoned us. We can use the predictions of this confounding theory with stunning accuracy in the tremendous colliders that particle physicists have used around the world. It is probabilistic instead of simply deterministic, but it is still predictive and not utterly wild and random.

Despite his profound contributions to the quantum model, Einstein never came to terms with quantum mechanics and declared that 'God does not play dice...'. To the last he believed that another theory would come along without the logical puzzles, without the ambiguity of probabilities. So far, a century later, no theory has made itself known, nor do we have a clear and universally accepted interpretation of quantum mechanics. The most widely referred to interpretation of quantum mechanics is the Copenhagen Interpretation, due largely to Niels Bohr and his discussions with Einstein, Werner Heisenberg and the suicidal Eherenfest. The Copenhagen Interpretation is implicit in modern quantum theory. The primary tenet of the interpretation is that the probabilistic nature of quantum mechanics is fundamental. For instance, the uncertainty principle is not due to an experimental lack of precision but instead results from a cardinal ambiguity. The particle does not have a definite location or a definite velocity but rather is in a superposition of possible states of being. Still, we don't despair of these ambiguities in interpretation because we continue to use the results of quantum mechanics with great success.

Quantum theory led to a modern theory of particle physics which distilled the laws of nature to four fundamental forces: gravitation, electromagnetism, the weak force and the strong force. The strong force binds together the quarks, which are the subatomic constituents of nuclear particles, while the weak force dictates the interactions of elusive particles such as neutrinos, which affect us so weakly that we are undisturbed and unaware of a storm of neutrinos which pass through our bodies and our planet as we speak. Many great minds were able to unify electricity and magnetism and the weak force together into one theory, the electroweak theory. Gravity remains the most stubborn, resisting unification in a quantum description, let alone a fully coherent single theory of everything. But that is the great and as yet unrealized goal, for Einstein and for those of us who were awakened by his call.

The most conspicuous arenas where gravity and quantum mechanics collide is in the earliest universe and in the harsh cores of black holes. On that remarkable terrain of curved spacetime I try to get my footing. Topology is beyond even Einstein's theory, but before I go beyond I have homage to pay to his unwanted children: black holes and the big bang.

# 7

## DEATH AND BLACK HOLES

**29 DECEMBER 1999**

I'm in line at the Russian consulate in London trying to get a visa. I have to part with my passport for a couple of weeks as they investigate my official invitation into the former Soviet Union. Relinquishing my passport to a fierce-eyed woman fills me with trepidation. She was impatient with me. Disgusted with my stupidity. 'You want your visa now' she tells me, '£120. You want it at 4.00, £80. For £20 you come back in a week.' In celebration of a streak of reckless decisions I agreed to come back in a week and pay £20. I was so relieved to have made that not very good decision that I abandoned the passport without resistance. I love that passport. It marks the start of a life I could never have foreseen and am not even sure I want. My old passport never looked like me. Issued in Boston, a city for which I have no intimate feelings, it's full of stamps. This passport was issued in London four months ago. I look even less like me. It has my British working papers and a wad of blank pages. This passport is all future, no past.

I'm on my way to Moscow for a meeting. My first trip there and I can't wait. All of those cold war movies unintentionally romanticized Russia.

When I get to Moscow, I'll be talking about chaos and the big bang. The meeting is in honour of Isaac Markovitch Khalatnikov. I have never met him but have studied his work. Neil Cornish and I worked on Khalatnikov's theories and were able to prove that his model of a generic big bang and the demise of a star to a black hole would be chaotic. I fear that work will be lost in obscurity, but I was impressed once again by nature yielding to mathematics. My interest in chaos was sparked by Neil and our collaborations. The topology of the universe is a virtually unrelated topic, although there was a tangential connection with the underlying mathematics of chaos and topology. Peculiar that Neil and I both worked on topology, somewhat independently, but we probably inspired each other's interest. These days we're both still working on black holes, general relativity's peculiar fledglings.

In its ultimate formulation, general relativity is just one mathematical sentence. It looks like: $\int d^4x \sqrt{-g}\,(R + L_m)$. This expression, known as an action, means that spacetime curves in response to the energy and the mass living on its three spatial dimensions and flowing through its one time dimension. From this one mathematical sentence people have been deriving unbelievable consequences, consequences that even Einstein often wanted not to believe. From the dynamics that precipitates from this law people were able to derive the evolution of the universe, the possibility of wormholes, black holes and the bending of light.

This one action, the Einstein–Hilbert action, named after you know who and David Hilbert (1862–1943), a spectacular mathematician responsible for the concise mathematical form of the action, can do all that. If you posit that the universe is full of a smooth distribution of matter, you can derive from Einstein's equations that the universe must be expanding and must have had a beginning, a big bang. If you focus instead on a very small region of the universe and consider what a massive lump like the sun does, you would find a curved spacetime that brings matter falling towards it or orbiting around it on precessing ellipses. If you make that lump dense enough, you get a black hole.

Everyone wants to know about black holes. It took decades for some very clever people to come to an understanding about black holes as a consequence of Einstein's theory. Schwarzschild died unexpectedly young, only after an incredibly productive year in which he rapidly isolated what would survive him as one of the most important solutions to Einstein's equations. Schwarzschild's outrageous solution describes the curvature of spacetime outside any sphere of mass and accounts for the motion around the sun as well as the earth. But the sun and the earth are also well described by Newtonian gravity. In other words, in weak gravitational settings, general relativity is very well approximated by Newtonian gravity and so the predictions are very similar. It was not until people began to pursue the implications of extremely dense stars that the real peculiarity of the Schwarzschild solution was appreciated. Near dense stars Newtonian gravity is a poor description and general relativity is needed.

Curved spacetime acts as a kind of lens redirecting light. Light moves

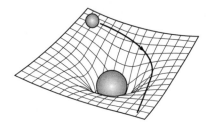

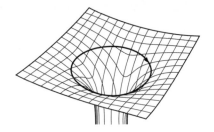

The bent path of light
around a star.

The circular orbit of light
around a black hole.

Figure 7.1 *The path of light bends as it follows the curves in space made by a central star. Around a black hole, light can execute a complete orbit, although it does so unstably.*

along the space on curved paths following the natural grooves. If the star is distended, like our own sun, light will bend as it passes the sun but will not bend enough to execute a full orbit (Figure 7.1). Around a compact, collapsed star the bending is more pronounced, but only the densest compact core can trap light in orbit. If enough mass is concentrated into a small enough region, the curvature of spacetime becomes so harsh that nothing can continue to orbit stably; not even light can escape this curved hole in space and the darkest corpse of all forms, the black hole.

There were fifty-five years between the theoretical discovery of black holes and their acceptance as a realistic astrophysical possibility. Even then acceptance came readily only from the theorists. Observational astronomers were sceptical for another fifteen years. It wasn't until 1967 that the American relativist John Archibald Wheeler coined the name 'black hole' despite French resistance due to the obscene connotations the name acquires under translation.

Now the astronomical community accepts that black holes are the final corpse from the catastrophic death of a massive star. Birth comes before death, so star formation seems a good place to start.

**30 DECEMBER 1999**
We really do look like a colony of ants strung in ragged lines going up escalators, down escalators, crushing into trains and spilling out of them like blood cells out of a broken vein. There I go, mixing my metaphors. Entomology or physiology, which do you prefer? Where is everyone going? I for one don't know. I'm momentarily disoriented. I'm in the

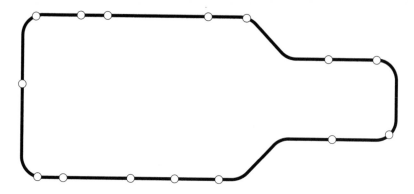

Figure 7.2 *The London Circle line.*

London tube, but I don't know if I'm coming or going. I should never have taken the Circle line. The Circle line (Figure 7.2) is the most heinous tube line in all of London. It's name is the most accurate thing about it, since the line does trace a topological circle, if not an actual circle, which has probably contributed to my disorientation. No beginning, no end. It gives no clues as to whether your stop is in front of you or behind you, since it is both.

Here we are drones in the vast circulatory system of the London Underground, shielded from the light of day. When I get out of here I will pay tribute to the sun, not staring at it the way Newton foolishly stared at the sun in his early experiments, no, I will instead accept a bit of warmth, a bit of vitamin D, a bit of sunshine. It is no wonder people have worshipped the sun throughout recorded history. Thousands of years ago the earliest Sumerian astronomers viewed the sun and the heavenly bodies as celestial gods. They carved tablets into visual tales of the living sun and his glorious companions.

The orientation of our astronomers has changed, but in a sense our sun is alive in as much as it still has fuel to burn. The sun will continue to live until the vast reserve of nuclear fuel is extinguished. The atomic energy is released through nuclear fusion, which begins with Einstein's most quotable formula: $E = mc^2$. Mass times the speed of light squared is equivalent to an energy. The conversion of mass into energy powers nuclear bombs and fuels the sun with a natural nuclear energy source. $E = mc^2$ is behind the most potent life force in the solar system and the most violent.

Einstein was plodding along in pursuit of something, call it truth for lack of a better word, and can't help that he recognized the latent power of nuclear energy. And now it's the stuff that bombs are made of. Hiroshima, Nagasaki, destruction, obliteration. Knowledge is power and power corrupts. Ironically, one of the greatest gifts of humankind, one of the greatest geniuses we have ever known, might inadvertently have taken one of the steps towards our demise.

Nuclear fusion requires extreme conditions, whether it be in the heart of a bomb or in the core of a star. The extreme conditions in the centre of the sun are provided by the pressure of gravitational collapse. Since every mass attracts every other mass, nebulous gas adrift in the galaxy can begin to clump around condensations of matter, eventually forming a star. The denser the star becomes, the hotter it becomes until it becomes so hot, unfathomable millions of degrees, that the atoms in the deepest interior collide with such forcefulness that their nuclei actually stick and merge. Through this nuclear fusion light elements bind to form heavier elements. Enormous amounts of energy are released in these nuclear furnaces and the star begins to shine.

Millions of tons of matter are converted into energy. Enough to keep the sun alive and shining for tens to thousands of billions of years. The X-rays and gamma-rays emitted in the nuclear reactions deep in the interior are absorbed and re-emitted by the stellar atmosphere. When the light finally emerges and travels through the solar system, impinging on the planets, most of the radiation is in the visible spectrum. By no random coincidence this is the light our eyes are tuned to and which has shaped our vision and perception of colour.

Yellow is in the middle of the visible waveband, is the colour of the sun and is also the colour used to code the Circle line on the famous London tube map. It is a topological map, which means the distances between stops are not accurately represented and only the connections are depicted. This Circle line train has pulled into South Kensington, a good stop. I'm going to get out here and visit the theory group at Imperial College. Must have been my intention all along. I'll visit Chris Isham, Jonathan Halliwell and Joao Magueijo. We'll spend a lot of time chatting and giggling before the day runs out and I follow my loop backwards through the tube.

## 31 DECEMBER 1999

The great Russian physicist Lev Davidovich Landau (1908–1968) is another great character. He tried to save himself from Stalin's purges, which had slaughtered millions and scorched Soviet intellectuals. The knock came to his door despite his attempts to protect himself with international acclaim, and they arrested this great talent that had sprung from their own soil, a patriot, a Marxist, a scholar, and they imprisoned him. For one year he survived his cell and the fate that had ushered in the near extinction of Soviet science, literature and art. His colleagues pleaded to Stalin himself for Landau's release. His sentence was commuted and he was freed, sickly and thin, and permanently marred by his incarceration. I think of him here because in a desperate attempt to evade Stalin's rampages he set forth ideas about the sun and the stars which did not stand the test of time. But the intrinsic elegance of his ideas has had a lasting impact on all of theoretical physics and he even made good with a Nobel prize-winning explanation of superfluidity, a new state of matter. But who can imagine what scars he kept covered?

Stars die when they run out of nuclear fuel. Without the energetic resistance of thermonuclear fusion, they can no longer support their own weight. They begin to fade and collapse. The death throes of stars take the form of white dwarfs, neutron stars or black holes.

If a star is as small as the sun, it will only swell and bloat with its coming death and eventually just die. These stars die in the form of white dwarfs. White dwarfs have a density of nearly sixty tons per square inch. The gravitational weight of the star crunches electrons into an alarmingly small area, much smaller than the electron would tend to occupy in the absence of the gravitational crunch. The electron rails against its confinement by vibrating quantum mechanically. This quantum pressure is a direct consequence of the Heisenberg uncertainty principle. Any attempt to force the electron to occupy a precise location results in a frenzied retaliation in velocity. The internal pressure to support the white dwarf against further gravitational collapse comes from this electron quantum pressure. White dwarfs shine for a while, emitting light from the motions of the vibrating electrons. Eventually this last gasp fades too and the white dwarf goes dark.

A bigger star will rebound as the matter begins to collapse with such violence that it produces a supernova explosion. A supernova casts off a staggeringly bright shell of the star's old stellar skin and atmosphere. Supernovae can shine a million times brighter than the progenitor star. They are rare enough that their appearance in our own galaxy is a spectacular astronomical event. Tycho Brahe, the guy without a nose, was able to see with his naked eyes a supernova appear in the sky and outshine Venus, which brightly decorates the evening sky. A few decades later Kepler and Galileo also documented the appearance of what we now think was a supernova. There is also some speculation that native Americans incorporated a bright object loitering near the moon in wall paintings and that this bright object may have been the same supernova that the Chinese documented in the eleventh century. The supernova remnant was bright enough to fight against the luminosity of the sun and be seen in the light of day. Supernovae gradually fade over months or years until they're too faint to still be seen by the naked eye. In 1987 an amateur astronomer saw a piercingly bright astronomical object suddenly appear that couldn't be seen the night before. The explosion happened thousands of years ago, but it took light all of that time to reach us and announce the demise of another star.

The remnant of the exploded star, sitting in the centre of the supernova shell, is a neutron star. The rebound from the supernova is catastrophic enough to force the elements themselves to implode, forming a giant nucleus. Neutrons, like electrons, will resist their confinement and vibrate quantum mechanically in rebellion. Just as quantum electron pressure supports white dwarfs, neutron stars are supported against further collapse by this neutron quantum pressure. A neutron star roughly one and a half times the mass of the sun would be around twenty kilometres across. Bear in mind that the sun is over a million kilometres across. Neutron stars are like giant astronomical nuclei.

Neutron stars were first discovered in the form of pulsars – astrophysical lighthouses beaming light out of magnetic poles by a poorly understood mechanism as the star spins thousands of times a second. Imagine that. A star the mass of the sun but smaller than Manhattan spinning thousands of times a second. The possibility was entertained that these precise lighthouses were not natural artefacts but were a coherent signal from alien life. We send beacons into space, messages in a bottle, to declare into the silence of mostly empty space a meek 'we are here' or maybe one day a 'we were here'. Human-made signals tend to be extremely regular in contrast to naturally made signals. For this reason, extremely regular patterns are used as code, a kind of universal symbol of civilization versus natural phenomenon. In essence, we are relying on math transcending language. Pulsars are such regular clocks that when the Irish astronomer Jocelyn Bell discovered the pulses, they were called little green men (LGM), only half jokingly. Instead, working at Cambridge on Anthony Hewish's team of radio astronomers, Bell had discovered highly magnetized neutron stars in the form of pulsars – an achievement for which Hewish received the Nobel prize.

Planets don't burn nuclear fuel. If the sun didn't shine, we'd be plunged into permanent darkness. Jupiter is almost big enough to burn but not quite. If it were a bit more massive, it would be able to ignite nuclear reactions and light up as a star. We'd have two stars in our solar system. As surely as night follows day might not be so sure. Day could wane to evening and the second star could rise, giving us more day. Maybe there'd be no night. Or our orbit around the two stars could be so complicated that an intricate variant of nights following nights and days would preside. We'd look different, we'd be different. We'd probably acquire vision in different wavebands, have different tides, different reproductive cycles, different astrological tales, different mythologies.

## 1 JANUARY 2000

We're visiting friends in California. We've been here a couple of weeks. My eyes still glow in orange from my dream. My side of the bed is hedged in by a pile of dirty clothes, the bed packed a little too near the windows. I usually trample on the garments, but today I decide to implement a new policy of tidiness. My first gesture is to avoid the filth. I have to climb over Warren, who wrestles me into the laundry. Our day starts again as the last ended.

Tripping over the books and into the bathroom, I scrub away sleep in the sink and sweep my hair under an old tube hat we bought at the pound shop in Brighton where, as you might have guessed, everything costs one pound. I showed a friend my gloves from the pound shop and he told me I got ripped off.

I make it down the street to a coffee shop. The grinder sputters and a girl drags her feet lazily to launch it back on course. She hums to herself, harmonies to the constant hum of the city. That inescapable sound is louder today and my attention drifts over to it. I feel the sound through my feet. It buoys me up, keeps me hovering above lifelessness. The people rush past the window like sparks. A blur on top of the persistent orange of my heavy eyelids. I'll finish my coffee and I'll go. I'll forget this scene.

The glass door of the building catches the light behind me as I get back and re-enacts a scene from this morning's dream. A tornado of glass flowers spiralled upwards as I fell and then the world inverted and my falling was flying.

I sit over a giant coffee, a mammoth bowl of caffeine, a mockery of the swirling espresso from one of Goddard's films. My coffee's not swirling and I'm not hearing voices whisper about existentialism or the cosmic void. I am reducing my anxiety to mundane questions. Small, seemingly concrete and simple facts. Stars are born and then they die. This I can handle.

Stars are born and then they die. The stars supply the universe with heavier elements, creating them through nuclear fusion in their interiors. The lightest elements are responsible for nearly all the mass in the first generation of stars. In the primordial universe there was essentially only hydrogen and helium with only trace amounts of heavier elements. There was no oxygen and no carbon, and so no water and no basis for organic life. The primordial universe was a sterile cauldron.

Much of the material synthesized in the centres of stars gets ejected back out into space when the star dies. Eventually new stars and planets can form from the star's scattered remains. This next generation of stars, and more importantly their satellite planets, can be made from heavier elements like carbon and oxygen. Plasma can form on the surfaces of the planets. Maybe on one with optimal conditions, complex molecules form and an inanimate broth waits for the sparks to generate organic life. *Voilà*. A few hundred million years later, Africa blooms and here we are.

We have the peculiar realization that we are made of the stuff of stars. I was sitting in the back of my first astronomy class at Columbia when I learned this. I like to remember it as a cold day, but I'm not certain it was. I snuck into the huge old lecture hall with its antiquated wooden chairs in the basement of Pupin Hall. The Manhattan project to build an atomic bomb was initiated here and you can feel the presence of that time, that generation of scientists, preserved in the odd accidental notches on the wooden frames to the old chalkboards. Their feet wore those marks in the steep steps connecting the sunken front of the hall to the elevated back, where I hide over the steam of my coffee in its 'I love NY' cup. A black-topped laboratory table occupies most of the lecturing space to remind us of the nature of our subject. Joe Patterson was lecturing and throwing candy bars at anyone awake enough to ask a question, trying, and for the most part managing, to capture the attention of the nonchalant teenage audience. All those years ago I learned from Joe that we are stars' debris and it still gives me pause. Not yet willing to remove my coat and take notes, I listened and he pierced my drowsiness with this one fact. Our bodies are made from elements synthesized in stars. These gloved hands of mine, these gloves, are reshaped atoms made only in the centres of stars. I admit I missed most of what he said before and most of what he said after, but I knew something had shifted for me. I still didn't know then that I'd want to be an astrophysicist. But a seed was planted; another seed, I should say, since it wasn't the first.

It's tempting to see the entire cosmos unfold like a giant factory driven towards the production of organic life – the cycle of life and death and regeneration. We play it out on our planet as death and decomposition nourish our soil and new life burgeons from the dirt. Our inanimate ancestors. Their animate descendants.

## 3 JANUARY 2000

It's always hard to come back. We've crossed the Atlantic to resume our bitter struggle between hell and something really good. I like it here, but then I have a life to return to. History depicts Cambridge of a few centuries ago as unsavoury and gloomy. Apparently the cramped filthy streets were full of thieves, murderers and prostitutes. Students were forbidden to fraternize with the locals until late Victorian times. 'Town versus gown' they called it then and still do now.

As we live here in this peculiar backdrop of Newton and England, history and science, my work is shifting to black holes. There's no causal connection between the backdrop and my shift in research – just incidental. Black holes are the third and most dramatic possible result of the death of a star. As a massive star runs out of energy to stay lit, the struggle against gravity is lost. A very massive star, more massive than twenty or so times the mass of the sun, will collapse down to a core smaller than a neutron star. The gravitational collapse is so savage that the matter crunches past the density of a giant nucleus, past any density anyone had ever anticipated, so that the ultimate fate of matter is to perish in a singularity, a region of infinite curvature where spacetime itself just ends. The bright star burns out, becoming a black vortex in spacetime that fades invisibly against the darkness of space. Light itself would eventually fall in if it ever veered too close to this gravitational siphon, this hole in spacetime. This dark, dead star could not emit light or reflect light. It would be black, a black hole.

Black holes are inescapable, hence their contribution to fear and fantasy. The gravitational tidal force near the surface of a black hole would pull on a space traveller's toes more strongly than their skull, stringing them out into a Giacometti sculpture and then a wisp of shredded matter. If you could survive in a protected rocket in a near orbit around the black hole, the bending of light would send distorted images of your own self along this orbit. Light reflected off the back of your head would travel full circle around the hole and greet your face. Any closer still and not even light could orbit. Everything falls forever into the impenetrable darkness.

The demarcation of no escape was eventually assigned the name 'horizon'. It's a good name, since in general the word 'horizon' indicates the farthest distance to which we can see and we cannot see anything closer to the hole than this horizon.

From our vantage point far from the hole, we'd watch the lethargy of the clocks of an observer hovering near the horizon. At the horizon, the clocks of our doomed companion appear to freeze and his time suspends. An ill-fated traveller might only notice his peril if he tried to turn around and realized he could not. Not all the energy in the world could power a reversal of his fate. Inevitably, he will fall into the brutal tidal stretch of the most extreme curvature and disappear down the throat of infinite curvature, the singularity. Singularities are bad and we on the outside cannot see them for the horizon. The only observers who could ever discover them by direct experience could never tell us on the outside of their definite existence. They'd take their knowledge to the grave.

Even though we can't see black holes naked, we can see them dressed with matter. Black holes power some of the most cataclysmic events in the universe and we can bear witness indirectly to their voracity. The black hole will cannibalize any companion star. Material will spill off the swollen companion and fall into orbit around the invisible black hole. The spinning central hole will twist the infalling matter into a giant orbiting disk until the neighbouring star is torn to nothing and all that remains is this bright hot disk. As boiling gas splatters and drags around the spinning space, it can heat to millions of degrees. Bright X-rays are emitted which we can see from earth. Stolen matter burns bright, exposing the clandestine black hole.

Some real surprises about black holes came observationally. Wildly active clouds were discovered but remained anathema. It wasn't until the 1920s with huge telescopes that observations pushed these clouds out of our own galaxy and established them as separate galaxies made of trillions of stars with outrageously dense cores. The central black hole devours the swarm of stars, lighting galactic cores spectacularly bright.

In the 1920s no one suspected the cores of galaxies were black holes. It took from 1915 until the 1960s before black holes were accepted among relativists as real possibilities for the end state of the collapse of a dying star. It was some time later before observational astronomers showed any interest in the wild speculations of Einstein's disciples. No one foresaw that black holes as extravagant as 100 million suns could burgeon from the cores of galaxies, but now we think they do reside there in nearly every galaxy including our own. Today we still don't understand galaxy formation well enough to know how black holes came to fill galaxy cores and many cosmologists are currently devoted to finding an answer to that puzzle.

Black holes at the centre of galaxies are millions of times the mass of our own sun and are concentrated in the deepest part of the galaxy. Swallowing entire stars whole, they grow and spin. In the most active violent galactic centres, quasars, the infalling matter is funnelled along the spinning poles into giant jets. The jets are an astounding hundreds of thousands of light-years across. Light that takes only eight minutes to make it to the sun, would take hundreds of thousands of years to traverse the lengths of the jets. Our own galaxy has mellowed over the course of time. There are no active jets in the Milky Way and we live in a kind of placid galactic suburbia. Although a central black hole very likely occupies the core of the Milky Way, we live cautiously far from this core in the galaxy's spiral arms.

There are still bare black holes out there that we can't see even indirectly but undoubtedly lurk undetected. Astronomers hope to build giant detectors on earth and in space over the next decade that will detect another remarkable prediction of Einstein's theories, gravitational waves. These waves of tidal forces emanate from the innermost regions of even the darkest black holes (Figure 7.3). As black holes spin and orbit compact companions, spacetime ripples in response. Gravitational waves will bring to us the cleanest snapshot of naked black holes, not dressed in accreting matter or disks, and give the most direct evidence of the most peculiar of nature's creations.

People argued back and forth about the reality of black holes, changing their minds and vigorously endorsing each new conviction. Einstein himself was convinced that they could not exist in nature and that some physical process must save the universe from their formation. Twenty-five years after Schwarzschild discovered his solution to the Einstein equations, the mathematical basis for black holes, Einstein still argued against their existence. We now think that no force in nature can resist the gravitational compression of a sufficiently massive dead star to hinder the formation of a black hole.

J. Robert Oppenheimer (1904–1967) and his collaborators were important in advancing the understanding of black holes as the end state

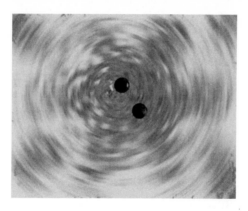

Figure 7.3 *Space ripples in response to the motion of matter and energy. The changes in the shape of space form gravity waves. Experiments are being built to try to detect these waves, particularly the ripples in space from the motion of two orbiting black holes.*

of gravitational collapse nearly thirty years before John Wheeler began calling them by that name. By the time other theorists began to accept his paradigm, Oppenheimer had been crushed in the stupidity of the cold war and betrayed by his country and even some of his peers. He was somewhat unmoved, it is reported, by his ultimate scientific vindication. I can picture the cool of his icy eyes. Was it resignation or had he just become so self-referential that not even the community's applause could affect him any more, let alone the community's reprisals?

**5 JANUARY 2000**
Black holes are simple. They always form so that, from the outside, they can be completely described by only three features: their mass, their spin and their electric charge. Unlike rocks, which are infinitely varied, each one unique, black holes cast off all variations, all perturbations, all imperfections to become all the same. Perfect, cold, dead stars, different only in their mass, their charge and their spin. That makes black holes extremely bare by nature. From outside the horizon, black holes are not differentiated by the original flow of gases, particle interactions or any other muck. They hide all of the information about their formation in their interiors. Their interiors are not theoretically understood and in any case no information about the interior can make its way past the horizon. We cannot see into the interior because light cannot escape. Nor can we receive messages from the interior, since no matter or

energy can escape. We have no way to learn about the matter and energy that was lost in the darkness of the black hole.

Still, there is something apocalyptic about black holes. In their interiors they betray the limitations of general relativity and quantum theory but keep the most extreme phenomena locked behind the impenetrable screen of the horizon. The dark exterior is just the surface. The cores of black holes are marked by a singularity: in other words, spacetime is infinitely curved down a nozzle in the core (Figure 7.4). Matter and energy fly towards a cut in the fabric of space. We cannot predict what happens to matter or energy as they wash down the sink, nor can we predict what might come flying out. What happens to matter and energy as they strike the singularity? Do they disappear out of the universe as we know it? Do they somehow escape oblivion? If matter can disappear down an infinite singularity, what is to stop new matter springing from the torn edge of space? General relativity does not help us answer these questions. We are left with no rules, no guidance, no means by which to predict what could happen at this singular core. Our knowledge of the laws of gravity fails at the edge of this rip in the universe. The infinite curvature associated with the central singularity is a Pandora's box of physical crises. Does nature really abhor an infinity?

Is relativity telling us that singularities simply cannot exist? Is general relativity announcing its failure to cope? Is there a theory beyond Einstein's that will avoid the ugliness of infinite singularities; a theory that can handle the extreme curvatures at the core of black holes without

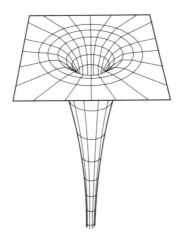

Figure 7.4 *A singularity is a region of infinite curvature. The heart of a black hole may be singular.*

becoming singular? And if so, what is this theory? We're inspired by the predictions of relativity to look for an even greater theory, a theory that looks like Newtonian gravity when gravity is weak and looks like general relativity when gravity is strong, but may look entirely different when gravity is strongest.

Where do we look for such a theory? The prevailing belief is that there is a quantum theory of gravity that will save us from infinities, and again relativity gives us clues about this ultimate theory, like a dying man scratching clues in the dirt.

The infinite curvature that relativity predicts acts as a magnifying glass pulling up to the surface all kinds of quantum phenomena. The hectic creation and annihilation of particles endorsed by the Heisenberg uncertainty principle is amplified by the extreme curvature of space. Space itself must participate in a frenzied fluctuating quantum foam. As the curvature gets stronger, the collusion of quantum theory and gravity is unavoidable. As quantum behaviour becomes more important, it is conceivable that quantum effects will avoid the formation of an infinite singularity altogether, saving us from unknowable tears in the fabric of space and thereby restoring predictability.

Quantum mechanics also has a remarkable consequence at the surface of a black hole. According to the Heisenberg uncertainty principle, a particle and antiparticle pair can be created from nothing provided they annihilate nearly instantly. Near the edge of a black hole, however, the particle can escape while the antiparticle falls into the hole. The black hole taps into quantum energy to radiate particles and lose energy in the process. Effectively, black holes evaporate. Stephen Hawking's proof of black hole evaporation is in part responsible for his fame.

The defining characteristic of black holes would seem to be their inability to emit light and matter. Quantum mechanics combined with general relativity alters the defining characteristic of black holes by allowing them to emit this Hawking radiation. The radiation is phenomenally faint, unobservably faint, for black holes we expect to populate the universe. Even though we can't really hope to ever observe Hawking radiation, the principle is of great theoretical importance. Black holes appear to radiate at a specific temperature related to the mass of a black hole. Bodies that radiate at a fixed temperature follow the laws of statistical mechanics: the description of the behaviour of small, atomic participants invented by Boltzmann. Like a giant natural allegory, black holes in their minimal form mimic the group behaviours of atoms. What this should have to do with gravity and curved spacetime makes the mind boggle. What are black holes trying to tell us?

Black holes are a portal to a fundamental theory of physics beyond Einstein, beyond quantum mechanics. Giving us these clues, black holes press us to pursue that final theory of quantum gravity. The other doorway to quantum gravity is the big bang.

# 8

## LIFE AND THE BIG BANG

In the 1920s the American astronomer Edwin Hubble observed that there were in fact galaxies beyond our own, confirming that the earth is not the whole universe, our solar system is not the whole universe, and neither is our galaxy. There are billions of stars in a galaxy and billions of galaxies we can see in the cosmos. From a handful of these galaxies Hubble was able to observe a chilling state of affairs – the universe is expanding.

Hubble deduced that distant galaxies move away from us. The further away they are, the faster they move. Confusingly, this gives an image of the earth at a centre away from which all galaxies race. We have long since dispelled the arrogant view that the earth is at the centre of the solar system and it would certainly be a moral regression to put the earth at the centre of the cosmos. The universe would have to look roughly similar even if we lived in a distant galaxy. If we lived in a distant galaxy, would we again observe all other galaxies receding from our new location? Is it possible to have all galaxies appear to recede from every other galaxy? The answer is yes, but only if we accept that it is not the galaxies that are moving but the space between them which is stretching (Figure 8.1).

For a visual analogy, imagine drawing dots on a balloon and then watching the balloon expand. Every dot will move away from every other dot, not because the ink has moved on the membrane but rather because the balloon is stretching between the spots. Similarly, imagine fixing all the galaxies on a three-dimensional grid and then stretching that grid in all directions. Every galaxy would move away from every other at a speed proportional to their relative separation. The further two galaxies are from each other, the more quickly they separate. The

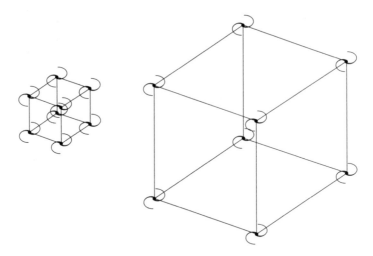

Figure 8.1 *Expanding three-dimensional volume.*

galaxies themselves need not move at all with respect to the grid. The stretching grid, which represents space, does all the work. For the most part, galaxies comove with the flow of the expansion, although there are deviations from this simplified picture. Some galaxies are bigger than others, and some clusters of galaxies are bigger than others. These dominant structures will pull on their subordinate neighbours, causing small local motions against the background expansion. These peculiar motions are generally less noticeable the further away the galaxies are from us.

If we trace the expanding universe backwards in time, the galaxies must have been closer together in the past. If we could run the history of time backwards, like rewinding a movie, we'd watch every galaxy move towards every other galaxy until they collide. The pressure of these collisions is fierce and if we could play this game of running the expansion backwards so all space shrinks, we'd have the inimical satisfaction of smashing all matter to bits. Galaxies would bust apart into their atomic constituents until the heat generated by the implosion split atoms into their subatomic particles, creating a fiery broth. If we go back in time far enough, the entire spacetime contracts to one point. All matter and space slammed together. The tepid name, 'the big bang' doesn't do justice to the inhospitable brew, the ultimate and most energetic merger of all matter, energy, space and time.

Running the movie of the history of time forwards again, the big bang issues forth a fiery ember in the birth of the universe. The big bang is not an explosion in space like a star exploding, where a ring of nebulous material surrounds an identifiable centre. Rather, the big bang is the creation of space itself, of time. There is no sense to the question: how long was it before the big bang happened? Time started with the big bang. There is no sense to the question: where did the big bang happen? It happened everywhere. The earth *is* at the centre in a sense. Every galaxy is at the centre. The centre is everywhere. The first sketch of our universe, its birth, life and death, begins here.

There was a Professor Segal at MIT who would always stop every visiting speaker whose talk had a title even peripherally related to cosmology. He'd then harangue the speaker in protest against their acceptance of the Hubble expansion. He had an alternative theory. I never did stop to find out about his theory or if it was at all consistent with general relativity. I don't really feel badly about that, even though I thought he seemed like a nice man from my seat way up in the back, where I tried to mind my own business. But who wouldn't want to believe in relativity? It's the best.

**7 JANUARY 2000**

Cambridge. I think Warren's having a nervous breakdown. I realized it was serious in the giant suburban Sainsbury's. We walked three miles to get there and were filling a cart with cans and economy bags of frosted flakes. Parents were whining, kids were crying, carts were crashing. In the middle of this suburban tornado of Sunday grocery shoppers, Warren was frozen. His face was stuck in an expression of disbelief, almost surprise. He was locked to that cart, rigid as stone. I saw all of this from the freezer section through a blur of shoppers. This was bad.

He follows me from grant to grant, city to city, different countries. He cooks, cleans, does my laundry, my shopping, organizes my bills, my life. He used to have his own life, his own ambitions. So many of my colleagues' wives have suffered the same trauma. None of us knows what to do to make it right.

He starts to tell me how he has to scrub the carpet and he'll do it when I'm at work, maybe Monday, it might take some time to dry, he could do it another day but this scrubber isn't good enough and he's unsatisfied with his cleaning products and he's yabbering, which in itself isn't that odd, but he's choking on his words and can't speak quickly enough. These products are no good, they've gone past due date, too much sugar, doesn't pour well, sticks to the can.

It's winter. Winter is always the worst. It'll take time, just enough time for the next move. We make it back to our rented house. We can barely remember where it is. He goes upstairs and plays the mandolin for a few hours and I finish a paper on black holes, chaos and gravity waves. I wonder why I keep doing this, how much longer we can survive. I've been here before, with others.

I guess we all resist change some time or another. Change is inevitable. Change and process will probably one day be all we consider in theoretical physics. We'll realize there are no particles, there's no spacetime, just the relations of events. Just change. But in the meantime we have to use the language of curved space and it has brought us epiphanies, like the realization that the universe expands.

Some find solace in the permanence of a static universe, an infinite cosmos that existed forever into the past and will exist forever into the future. Even Einstein, who willingly accepted the impermanence of time and dethroned spacetime from absolutism, still resisted the idea that the universe should expand. There he goes, resisting change. Einstein actually altered his original theory by adding a now famous term called the cosmological constant. With the cosmological constant his equations predict a static universe, though only precariously static since a slight perturbation would send the universe expanding or contracting again. But Edwin Hubble's observations of the recession of galaxies confirmed a cosmic swelling. The expansion is exactly consistent with an expanding space. Einstein capitulated with some self-reprobation.

Einstein resisted the expansion, Woody Allen panicked over the expansion. Which movie is it in which Woody Allen refuses to do his homework because the universe is expanding? His mother brays something to the effect of 'What business is it of yours? You live in Brooklyn, Brooklyn is not expanding.' The motion of space fills some of us with anxiety over our mortality. If the universe had a beginning, it will have an end. If the universe had a birth, it will have a death. Our origins and our terminus can both be predicted from Einstein's theory.

Alexander Friedman had more faith in Einstein's theory than its inventor. While Einstein mistakenly rejected an evolving universe, Friedman predicted expansion from Einstein's own equations. Friedman's prediction of an expanding space predates Hubble's observations. His ideas languished in obscurity for some time until Howard Robertson and Arthur Walker developed similar models to reconcile Hubble's observations. The Friedman–Robertson–Walker solution still forms the primary background against which all of modern cosmology is formulated. Some people remember to include a Belgian priest and scientist in this triple-barrelled tribute, Georges Lemaître, who also independently understood the Friedman–Robertson–Walker solution. Some are given to calling it the Friedman–Robertson–Walker–Lemaître solution. A Russian, an American, a Briton and a Belgian.

Friedman found his solution to Einstein's equations by taking the Copernican revolution further to argue that we should not be special in the cosmos. As the universe looks to us, so it should look at any other location in space and in any other direction, at least on average. The particulars may be different, but roughly the cosmic conditions should be the same. This cosmological principle amounts to imposing two symmetries on Einstein's equations: homogeneity and isotropy. Homogeneity asserts that the universe is the same in all locations and isotropy asserts that it is the same in all directions. With these two symmetries Friedman was able to simplify Einstein's equations and derive a description of the evolution of the universe. Actually he found only one solution, although there are three solutions consistent with homogeneity and isotropy.

Here's how it works. Space warps in response to mass and energy. If the universe is full of a relatively smooth and evenly distributed amount of matter, it will curve and expand in one of three ways and our universe appears to be among these three options. The space is either flat, positively curved or negatively curved. These are the three-dimensional analogues of the curved surfaces in Figure 5.2. All three Friedman–Robertson–Walker–Lemaître spacetimes experience a similar birth and early history, but all have different destinies.

The solution Friedman found corresponds to a space with positive curvature. If there is enough mass and energy to overcome the cosmic expansion, space will eventually begin to contract. The curvature of space is everywhere positive, which makes this overdense universe a three-dimensional generalization of the earth's two-dimensional spherical surface, known as a 3-sphere. The 3-sphere is always finite, just as the earth is finite, and we have our first mathematical prediction for a finite universe.

If there is not enough matter in the universe to overcome the cosmic expansion, space will expand forever. This underdense universe is everywhere negatively curved like a saddle of sorts. The negatively curved space is infinite in its simplest incarnation, but I will go to some trouble to convince you later that it need not be infinite.

The third and final geometry marks the critical balance between the other two. The matter and cosmic expansion are exactly balanced and the universe expands forever, approaching a static universe in the infinite future. The curvature of the critically dense space is zero and so it is everywhere flat. Flat space can also be infinite, although I hope to show you that flat space can be rendered finite.

We don't know which one we live in yet. Will the universe expand forever and die cold and sparse, or will it one day recollapse, long after our own sun has expired, bringing cosmic extinction in the form of a relived big bang? We don't know yet, but we almost know. We live in an age when we can very nearly answer this question. We live at a time when people have the tenacity to analyse the data of telescopes, build satellites and develop detectors that, when put together, could actually determine the fate of the whole cosmos.


## 11 JANUARY 2000

My generation grew up after a man had walked on the moon, Apollo 1 had fallen to tragedy and Apollo 13 had defied tragedy. I knew about supernovae and stellar evolution, galaxies and space. I was enthusiastic about NASA, the future and space travel, but it was no longer sheer fantasy. It was all possible. We weren't jaded, but maybe we took for granted that nature was knowable and accessible. Then there are times when nature alludes to our place in her larger scheme.

Dad and I used to run along the beach at night in Hilton Head Island, South Carolina, and when we had clocked in our miles we'd collapse into the sand along some fallen reed. But it wasn't the ocean we'd look at. With no street lights on the island, the ground was seeped in a blue-black darkness, obscuring the beach, which we could hear but couldn't see. There it was, the Milky Way. I can't remember the first time I saw that milky rift of stars cutting through the sky. I know it was on the island that I saw it in its real splendour, but time stood still for me while I looked agog and each time it's like the first time, so they all bleed into a timeless picture of that edge-on view of our galaxy. 'That's our galaxy,' I'd say to myself with a peculiar combination of belief and disbelief, as though for the first time I understood my childhood lessons. The ground dark, the sky so bright you would choke with surprise, I could virtually feel the rolling of the earth under this blanket of stars. Dad would ask me what little I knew about astronomy and I'd babble, because I tended to get excited and talk too much but this was one of those occasions when he didn't mind. We saw remarkable things those nights, haunting images seared in my mind from shooting stars to lights so strange and random that I didn't even try to guess what they might be.

Now I know simultaneously more and less. Each answer I learn releases a rainfall of new questions. While we try to determine the nature of our ultimate end, we can decipher our common beginnings. With general relativity we have a scientific theory of genesis. Here is how the story of our early universe goes. In the beginning there was nothing. And I kid you not when I say nothing. I mean nada, zero, zip. No space, no time, no matter. Nothing.

From nothing, a universe generates spontaneously. According to the classical precepts of relativity, the big bang initiates a singular point – infinite curvature and infinite energy scales. Singularities are very, very bad things and are so disturbing that we generally regard them as a mark of the failure of Einstein's theory and not a real occurrence. This is not to say that there was no big bang, but that we are safest trusting Einstein's predictions only for times a fraction of a second after the bang when quantum effects have died down.

Isaac Khalatnikov and Evgeny Lifshitz suggested that the predicted singularity was a consequence of an unrealistically symmetric model of the big bang. They argued that a less idealized model would not require the mathematical singularity and would therefore be a more wholesome theory of our origins. In fact the natural exit from a big bang might not be symmetric at all (Figure 8.2). The simplifying assumptions of homogeneity and isotropy lead to a picture of the universe that expands at the same rate in all directions. Khalatnikov and Lifshitz described a more general possibility of a universe that expands at different rates in each direction. The space would contract in some directions and expand in others, so that the whole universe alternated in an unpredictable sequence of these oscillations.

The debate that followed saw the blossoming of Roger Penrose's great

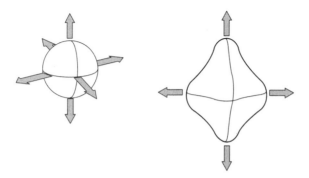

Figure 8.2 *Isotropic versus anisotropic expansion.*

talent and his mathematical methods – topological methods actually. Penrose was able to show that black holes would always harbour singularities and Hawking would extend those results to prove that the big bang would always initiate from one. In sum, their proofs constitute the Hawking–Penrose singularity theorems.

Lifshitz and Khalatnikov, working with Vladimir Belinsky, would later concede that the singularity was unavoidable, but they nonetheless made an important point about the erratic behaviour of the early universe. They argued that if the universe began inhomogeneously and anisotropically, it would behave so erratically as to be truly chaotic. Chaos is an extremely elusive term. There are several formal ways to define chaos, but it's not clear that they all agree. Intuitively, chaos refers to a system of such extreme sensitivity that it becomes impossible in practice to predict the outcome. In the case of the Belinsky–Khalatnikov–Lifshitz model of the big bang, if the universe began even slightly differently – minutely different in the energy or in the initial expansion – then the entire history of space and its oscillations between expansion and contraction would be different. The universe is still deterministic, so that a computer could evolve the Einstein equations from a specific initial state for the universe to a determined fate. However, if the initial state is even slightly different, the entire evolution of the cosmos will be altered. Since we can never specify any state so precisely, we cannot ever predict how a chaotic universe will unfold. Eventually the chaotic episode will end as matter and radiation become dominant and dictate the course of the cosmic evolution.

When John Barrow was a postdoc at Berkeley, he worked through an elegant description of chaos in the big bang. One of the more agonizing aspects of this subject is how slippery conclusions can be. Barrow's results seemed unambiguous, but the relativity of space and time renders all claims about chaos relative as well. What may look like an extreme sensitivity to one observer may look quite moderate to another. It became difficult to assert with confidence that chaos and disorder truly prevailed. Nearly twenty years of debate on chaos in curved space followed. A few years ago Neil Cornish and I found a way to resolve the debate and show that an asymmetric universe would in fact develop chaotically. Oddly enough, Neil and I used topological methods also, but of a very different variety. We exploited modern chaotic methods and the power of fractals. Fractals are spectacular geometric objects which repeat their structure on smaller and smaller scales (Figure 8.3). The closer you look at the segments of a fractal, the more structure

Figure 8.3 *An example of a fractal; a geometric structure which repeats its patterns on ever smaller scales.*

you'll find. The snowflake is a fractal, the coastline of England is a fractal, some of Cantor's infinite sets form fractals.

The rest of the astronomical community care little about our results. Most of astronomy pursues more concrete gratification. Quantitative and definitive results form the bulk of astronomical knowledge; measurements of the temperatures of stars, determinations of their masses – concrete observations. Our more abstract pursuits are seen as too fugitive to be satisfying. Understandably, many prefer the satisfaction of the concrete to the potential emptiness of chasing apparitions. I find the abstract more than satisfying. It makes my knees weak, as though I've been privy to a few of nature's secret messages.

**13 JANUARY 2000**

Why does a universe spontaneously appear out of nothing? That's still dodgy territory and more the realm of a theory beyond Einstein's, a theory of quantum gravity.

I'm sitting over a curry with Lee Smolin and Lisa Randall. Lee's an expert on quantum gravity. Lisa's an expert on particle physics. We are speaking excitedly and probably loudly. If we're drawing attention, we don't notice. Today we all feel really confident about how much we don't know. 'Stupid question...' I announce to prepare them for what's next. 'Do you all believe in a big bang?' 'What do you mean by a big bang?' Lisa asks and off we go. Do you think the universe existed infinite and forever or do you think there was once nothing and now there is something? They concur with some degree of revulsion that both premises are totally ridiculous. Well the universe *is* expanding, we all agree. Neither of them will risk conjectures on what might have come before the quantum era. Fair enough, but I press and they wince. We just don't know, they concede, and though I think them both daring and innovative, I wonder if they are seeking shelter in safety or are implying something profound about meaning and about knowledge and I'm being stubborn. Someone intimates that maybe it's not a scientific question to ask until we know more about the theory of quantum gravity. Hmph, I declare with disappointment, but I don't persist.

I've known Lisa for eight years. She was a professor at MIT when I was a graduate student. We've never spoken about one very obvious commonality. We've never so much as alluded to our gender. The subject comes up today and we groan with an element of relief and a mutual eye rolling meant to indicate a sense of 'if you only knew' or a 'you can't even imagine', only we both do know and can imagine. I don't ever talk about that any more, I tell her. She's just starting to talk about it, she confides in me. We laugh with appreciation at each other's progress. I leave her in the tube on her way to Heathrow and back to Boston. This time I think we'll stay in touch.

Appease yourself with what we do know with fair confidence. Just after the Planck time, as it is known, quantum gravity has yielded power to an era that is fairly well explained as a cosmology bound by the dictates of general relativity with some quantum fields living in it. The Planck time is a natural time scale associated with the strength of gravity and is related to Newton's constant, from Newton's original description of the force of the earth pulling on the apple in the tree. Quantitatively, the Planck time is extremely small, specifically $10^{-43}$ seconds (that is, 0.000...1, with forty-two zeros between the 1 and the decimal point) after the birth of the universe. But some flash of a moment later, around $10^{-35}$ seconds (that is, 0.000...1, with thirty-four zeros between the 1 and the decimal point) after inception, space was filled with a primordial soup of light, quarks, electrons and neutrinos. The list of ingredients is a list of the fundamental constituents of matter.

Rocky Kolb, a cosmologist from the University of Chicago, used to hand out cans of soup with a twist on Andy Warhol's twist on Campbell's soup. The labels read 'Primordial Soup'. The primordial soup he gave me had listed as its ingredients things like: chicken, water, salt, sugar and then some synthesized chemical compounds. Needless to say, there was no chicken in the real primordial soup. There was no water, no salt, no sugar. If we could create a universe in the laboratory and were to drop a pinch of salt in the boiling soup, it would immediately be bombarded by the energetic frenzy around it and break up into its fundamental constituents of elementary particles. Of course, this is a nonsensical experiment and luckily we can't create a universe in the laboratory, or who knows what havoc a reckless engineer could wreak? The universe at its inception was smaller than a grain of salt and unfathomably dense.

Light and all of the other elementary particles collide feverishly, so that the cosmos is filled with a blinding storm. We live in an atmosphere that is essentially transparent to visible light. We can see across a landscape because of the air's transparency. In the early universe, the cosmic soup was opaque to radiation, which is to say that light could not travel freely across space but instead was scattered off hot particles. At these high energies, light acts particle-like with the individual photons scattering randomly through the buffeting primordial plasma. The collection of photons is in equilibrium with the other ingredients in the soup and they have the same temperature, since temperature is related to the average energy of motion of a collection of particles. If we had eyes tuned to see such high-energy radiation, we wouldn't be able to see anything but the thick frenzy of radiation. As the universe expands, it cools and settles and calms down a bit. Eventually, the universe cools enough that the motion of particles slows and their collisions are less energetic. The fundamental elements are able to settle into primordial nuclei without breaking apart. The gas formed is primarily positively charged hydrogen nuclei and negatively charged free electrons. The soup is still hot enough to keep light scattering off these charged particles and remains an opaque, hot bath at temperatures over billions of degrees. After about one second until about three minutes into the life of the cosmos, there is a period of primordial nuclear fusion when helium forms and trace amounts of light metals like lithium form. Still no oxygen, no carbon and so no water. Still no chickens.

Within these first three minutes the universe rages through a dramatic and rapid development. A minute is a natural unit of time for us on earth, who measure years in terms of orbits around the sun, days by the earth's rotation and hours by the sun's natural progress around the hemispherical sky. Each tiny fraction of a second, the nascent cosmos leapt forward in its development. After three minutes not much happens, at least not so quickly.

After about 300,000 years the universe cools enough that collisions with matter become less frequent and light is not so effectively scattered. The cosmic soup becomes transparent to light, allowing radiation to pass freely through space. This time is known as the time of last scattering and marks the transition from an opaque to a transparent cosmos as the blinding storm gives way to a clear path for light. This light from the time of last scattering travels essentially unimpeded throughout the universe for the next 10 or 15 billion or so years. In the meantime, the universe continues to expand and cool, matter begins to clump and bind.

Galaxies form, stars form. They live, explode, die. New generations of stars form, planets form. The earth is made, dinosaurs roam for millions of years, chickens finally make their appearance. Before or after their eggs, hard to say. Then humans arrive and after about 40,000 years we begin to build phones and rockets and telescopes. All the while we are submersed in this bath of background radiation left over from the big bang.

By now the temperature of the cosmic background radiation has cooled to a few degrees above absolute zero. Absolute zero is defined as the temperature at which a gas would have no pressure and is equal to −273.16 °C, which corresponds to −459.36 °F. This puts the wavelength of light in the microwave band. Everyone in the know just calls it the CMB for Cosmic Microwave Background. Physicists are obsessed with acronyms: WIMPs, MACHOs, POTENT, COMBAT. All allegedly stolen from the first letters of a meaningful phrase like: Weakly Interacting Massive Particles (WIMPs) or Massive Astrophysical Compact Halo Objects (MACHOs). The acronyms fit comfortably into a lexicon replete with terms like 'sterile', 'impotent', 'the bulge' and 'barrier pene-tration'. I saw Rocky give a talk once where he swore the objects he studied were Not Astrophysical Compact Halo Objects (NACHOs). At least his acronym had to do with junk food and not male insecurity. There are TOEs (Theories of Everything) and there's SUSY (SUperSYmmetry) where we start to get into the suspect territory of using more than first letters in the acronym, and lately people will use any letter in the phrase they like, which I think is strictly cheating. But the acronym CMB has stuck for decades and at least we remember what it stands for, unlike M-theory, where absolutely no one knows what the M is for.

## 17 JANUARY 2000

In the late 1940s a group of very ingenious physicists predicted that the universe should still contain a cool echo of the origin of the universe in the form of a cosmic background radiation. They even predicted the temperature and, while they got the exact number wrong, they were remarkably close. Most people thought they were mad. Again with the madness. George Gamow and his collaborator students Ralph Alpher and Robert Hermann were the first to put forth this suggestion in 1948. Another tale that gets passed down through the generations is that Gamow wanted to add a famous nuclear astrophysicist, Hans Bethe, to the author list so that it would read: Alpher, Bethe, Gamow – a pun on the Greek alphabet: alpha, beta, gamma. Hermann would be added to the author list only if he agreed to change his name to Delter in deference to the fourth letter in the Greek alphabet, delta. I suspect he refused.

In the 1960s other groups independently advanced the same suggestion as Gamow, Alpher and Hermann. Robert Dicke and James Peebles of Princeton University argued that there must be a sea of primordial radiation, as did Andrei Doroshkevich and Igor Novikov from the former Soviet Union. Then came a fortuitous but totally momentous discovery. Our civilization first measured the echo of the big bang from New Jersey.

Bell Telephone labs, in an attempt to improve telephone communication, sent the radio engineers Arno Penzias and Robert Wilson to test a huge radio antenna in New Jersey in 1965. The detector pointed at the sky and made sensitive measurements of light with such long wavelengths that it entered the microwave bandwidth, well beyond the perception of the human eye. Try as they might they could not get rid of a nagging kind of static in the antennae. They tried clearing away any possible interference including bird droppings. But it didn't go away. There it was, a quiet hum in every direction in the sky. Regardless of the tilt of the earth, the time of year, night or day, there it was, the same in all directions. The signal must have a cosmological origin beyond our planet and our solar system and even beyond our whole galaxy. A note sustained for billions of years, an echo of the big bang. Only they didn't know this was what they had detected.

The paper they published, which would earn them a Nobel prize, closed with the most humble suggestion that there might be a cosmological origin to the interference. I doubt they even believed that. It was too impossible, too much to ask for, too much to be given when they didn't even ask for it. Gamow and company must have been beside themselves. What a prediction to have made. Those millions of years of evolution, civilization, the accident of each of our individual births, to be born at the right time, in the right place, to be here after Einstein, before the discovery, to have been the ones to make the prediction. And to have a sense of humour too.

## 19 JANUARY 2000

Sometimes the beauty of San Francisco gives way to muck. I love

Chinatown but it is engulfed by an urban grit and stench. I wander through Chinatown when I go back to California, taking poorly chosen routes. The people spill on to the street, having run out of room on the sidewalk lined by vegetable stands. I take a lot of extremely amateur photos of curbs, manholes and graffiti, and survey them now. I have pictures of modified Chinese architecture miniaturized to fit on the San Francisco streets and fronted by distinctly American car metres, concrete roads and steaming manhole covers. It's hard for me to place the significance of culture and humanity in a universe that barrels along without concern for our welfare. Our city monuments are poignant, but maybe I don't know how to assess our significance. Why are we all struggling so desperately to survive? I don't know how to place us in the greater scheme of things. As Oscar Wilde said, 'We are all in the gutter, but some of us are looking at the stars.'

Sometimes I'm looking at the gutter and sometimes the stars. But when we (the collective we) do look at the sky, what we see is tremendous. The cosmic background radiation is one of those remarkable discoveries that will mark the century. The last century that is. In the early 1990s the COBE satellite (another acronym: COsmic Background Explorer) launched into orbit around the earth. For a few years the detectors on the satellite pointed out into the sky and measured the temperature of the cosmic background radiation in every direction. The early COBE data were first presented at an astronomy meeting to spontaneous applause. I have never before witnessed spontaneous applause at an astronomy meeting. The results were impressive. The cosmic background radiation does in fact fill the sky and on average appears to be completely thermal: that is, it can be completely characterized by the temperature alone. The cosmic background radiation looks exactly like a hot bath of light left over from the big bang. Only it's not so hot any more, just a few degrees above absolute zero.

After the COBE satellite a generation of cosmologists became slaves to the cosmic background radiation. Superb experiments are built to perform measurements of all kinds. Some are flown on high-altitude balloons and others are based at the South Pole. We're still waiting for the future and possibly final generation of cosmic background radiation satellites. One of these is an American project called MAP (Microwave Anisotropy Probe) and another is a European project called *Planck Surveyor*. The satellites aim to resolve fine images of the primordial light.

The cosmic background radiation is impressively smooth, having a virtually identical temperature in every direction. The nearly impeccable cosmic background radiation offers stunning confirmation that as far as we can see the cosmos is homogeneous and isotropic. If anything, it is hard to foresee how the universe could survive so very homogeneous and isotropic.

Friedmann made some sweeping simplifications when he assumed the entire universe was homogeneous and isotropic. And in fact, there must be tiny ripples in spacetime, tiny hills and valleys. According to the Heisenberg uncertainty principle, quantum particles and fields cannot be perfectly still. A consequence is that there will always be quantum fluctuations where the local quantum energy will be slightly higher or slightly lower than average. Space curves in response to the changes in density creating tiny peaks and troughs on the otherwise smooth space. Light will lose energy as it climbs out of valleys and gain energy as it rolls down hills. The overall effect is a patterning of the cosmic background radiation with minute hot and cold spots (Figure 8.4). These spots are very faint. By hot I mean that the radiation will be slightly higher in energy and shorter in wavelength, so that hot spots are slightly more blue than the average background light. By cold I mean that the radiation will be slightly lower in energy and longer in wavelength, so that cold spots are slightly more red than average. The hottest hot spot is only 1 part in 100,000 times hotter than the average and the coldest cold spot is only 1 part in 100,000 times colder than the average. The future satellite missions will detect these fine distinctions to build more

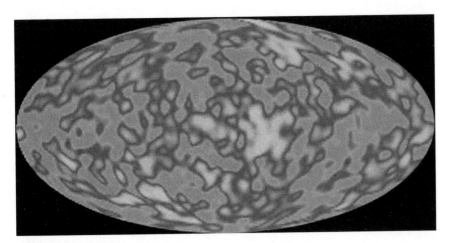

Figure 8.4 An elliptical projection of the hot and cold spots in the microwave radiation filling space, as seen from the COBE satellite. Taken from http://space.gsfc.nasa.gov/astro/cobe.

detailed maps of the patterns in the cosmic background radiation. I'm impressed with the bravery of the experimental teams that have managed such sensitive measurements. As tiny as they are, these hot and cold spots encode some profound secrets about the cosmos.

This archaeological remnant of the big bang has journeyed from the farthest reaches of the cosmos that we can access and carries information about these earliest times, and so encodes all kinds of information about the large-scale landscape of the universe. In particular, we should be able to see an imprint of the geometry of space in the pattern of hot and cold spots in the sky.

We can see how the universe got its spots.

### 23 JANUARY 2000

Einstein was gifted. Even his mistakes were strokes of genius. The cosmological constant that he invented to stop the expansion and that he later discarded to let the expansion run wild was revived to great effect in 1981 when Alan Guth realized that some subtle inadequacies of the standard big bang model could be gracefully erased by the reintroduction of the cosmological constant.

The gnawing question recurs, how did a universe that began so chaotically end up so smooth, so homogeneous and isotropic? After all, Belinsky, Khalatnikov and Lifshitz showed that if we begin in a universe with random directions expanding and contracting, the exit from the big bang would be chaotic with space churning unpredictably in alternating states of collapse and expansion. Surely such a violent beginning to the entire cosmos would come with scars on this perfect veneer. Yet, it does appear to be homogeneous and isotropic, since the temperature of the cosmic background radiation differs by only 0.00001 across regions of space separated by some 30-odd billion light years. Clusters of galaxies also seem pretty evenly distributed on the largest scales. The variations are disconcertingly small. The cosmic background radiation formed when the universe was still young, only a few hundred thousand years old. Since no information can travel faster than the speed of light, two points separated by any distance greater than a few hundred thousand light years could not possibly have been in causal contact. How did regions of space separated by nearly 30 billion light years come to have such precisely similar conditions?

If an anthropologist were to discover that two ancient civilizations had identical languages with only one word in a hundred thousand distinguished, they'd surely argue for a causal explanation. The two civilizations must have been in contact and the scientific task would be to explain how they had communicated. Guth's inflationary theory does this for cosmology. Guth suggested that there was an episode in the very early history of our universe, about $10^{-35}$ seconds after the big bang, when spacetime expanded at such a furious pace that a tiny patch was blown up large enough to encompass everything we can see today. The patch was small enough that it was within the maximum distance light could travel in $10^{-35}$ seconds. Inflation suggests that the universe has such uniform conditions today because everything could interact on these small scales and even out prior to the onset of inflation. The global implication is that the universe is not everywhere smooth; it is only smooth on this very small patch to which we bear witness. Imagine standing on a lumpy mountainous peak. You would certainly be aware of the treachery of irregularity. But if the small region around you were stretched as big as the earth, it would seem locally quite smooth to you, the variations being on such huge scales that you simply couldn't perceive them (Figure 8.5).

Another consequence of inflation is that we expect the universe to appear flat. If we stood on a basketball, we'd be aware it was round. But if we inflated that basketball to the size of the earth, we might be deceived into believing it was flat. The argument is not that the universe is truly flat but rather that we can't see the curves.

There have been thousands of papers written on Guth's inflationary theory since its proposal, clearly the biggest idea in cosmology for

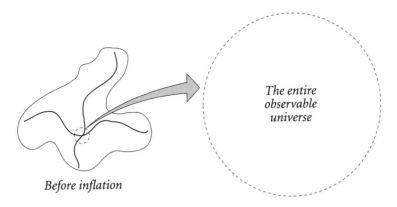

*Before inflation*

*The entire observable universe*

Figure 8.5 *During inflation a small patch on a lumpy space is blown up large enough to envelop the entire observable universe, large enough that the universe appears flat and smooth.*

decades. We still don't know for sure if inflation is correct, but it does well in the face of experiment. The problem is that there is no real theory of inflation. Inflation is more an idea that can be tested indirectly by determining if the universe is flat and confirming if the detailed marks on the cosmic background radiation match the inflationary predictions. Still, the engine, the driving force, remains unspoken for.

The inflationary paradigm offers a plausible explanation for how a lumpy, chaotic universe could end up vast and smooth. It gives us a chance of survival, since the universe must be vast and smooth to be inhabitable. Fifteen billion years later, we're here. Still, I do wonder if we haven't squandered that chance of survival. The dinosaurs managed to roam for 250 million years. All we have under our belt are a few tens of thousands years, an opposable thumb, some fire and we've already nearly demolished the planet. Maybe we're not such a success as a species if you measure success by the likelihood of survival. Even if we do survive, we have to at least admit that it's *possible* we will not and that our demise could come tomorrow. We could poison ourselves, toxify the earth, drop a weapon of mass destruction. Are we a suicidal species? Will we be responsible for our own genocide? There I go, obsessing about our predilection for insanity.

Yet we can ask these questions: why? how? We can even answer them. The dinosaurs couldn't do that.


**28 JANUARY 2000**

I admit I'm writing some of this months after the fact and my mind is defensively loosening its grip on that time. I remember meeting Warren halfway in the diagonal park where our move to Cambridge was decided months earlier. We're bundled against the wind and the rain, and I can see beneath the layers of cloth that his eyes are laughing.

We hug clumsily, our limbs thickened with layers of wool. We try to hold each other and kiss and then run home. At least I run and he calls me a fool, trying to catch up to me. Warren stammers and yabbers over his plans. He's now a self-taught luthier, making wooden instruments that unexpectedly materialize in our living room. His last American mandolin has dark wooden tones in both colour and sound. He comes in from the back room covered with wood chips and showing off his latest carving.

It feels like we're doing better, but I know we won't stay. I don't think we'll be here six months. It's too hard on him and it would be impossible without him. Our planning begins again. We can move to London and I'll commute to Cambridge. We can move back to Brighton and I'll move my grant. We can go back to California, but I'll have no job there.

Somehow my work is going well. I've felt welcomed in England. I'm impressed with how graciously and eagerly people discuss and banter and play with ideas. I am very fond of my colleagues here. But if we have to leave, I'm braced. I'm learning to make sacrifices that I was never able to make in the past and I paid for that inflexibility. Nearly eight years I spent with Andy and not a day goes by when I don't think of him. The pretty bracelet he slipped around my wrist is still there – a shackle or maybe just a reminder of my choices and my shifting priorities. For once I'm not going to put my work first, I lie to myself.

Seems easier to think about infinity. Seems easier to explain infinity to my silent, absent audience. When I concentrate on math, the stress of the present wanes and is replaced by the uncomfortable paradoxes surrounding infinity. This much seems clear: the universe appears to have one of three geometries. Space may be positively curved due to an overdensity of matter that will one day cause the universe to recollapse, or it may be negatively curved due to an underdensity of matter that will condemn the universe to expand for eternity, or it may be flat and of critical density so that the universe will slow to almost no expansion. Each of these geometries has a simplest topology, a simplest possible global shape: the simplest positively curved space is finite, a three-dimensional generalization of the surface of a sphere; the simplest negatively curved space is infinite and edgeless, as is the simplest flat space. In the standard model of cosmology, it is assumed that the universe is born with the simplest topology and so is infinite if it is flat or negatively curved. The positively curved space is born finite, but no one really believes that the universe is overdense simply because cosmological observations do not seem to reflect an overdense universe. If anything, the universe is underdense or critical according to the astronomers who have taken the time to lift their heads off the page and look and see.

According to the standard assumptions made in cosmology, nothing came from something and an infinite amount of something. When the universe is born with an infinite topology it is born instantaneously infinitely big. No information is communicated faster than the speed of light, but two points on the surface of space are ripped apart from each other faster than light's speed, so they are taken out of communication, never to be rejoined since it would take an infinitely long time for them to ever come into contact.

So is that it then, is that the end of our journey? Is the universe finite only if it is positively curved? Is the universe infinite if it is flat or negatively curved? The answer is no. That's not it. The reason the answer is no is that general relativity is not the whole story.

# 9

## BEYOND EINSTEIN

**31 JANUARY 2000**

Einstein was eminently quotable: 'Only two things are infinite, the universe and human stupidity, and I'm not sure about the former.' Neither am I sure about the former.

Einstein's revolution remains incomplete. General relativity signals its own failure. The Hawking–Penrose singularity theorems prove that essentially all big bang models and all black holes will possess singularities, although a singularity has never been directly observed. Singularities are regions of infinite curvature. They mark an unholy end to spacetime where matter and energy just terminate. We cannot predict the fate of a path that ends there and strict determinism is lost. Information, matter and energy can fly in and out of a singularity with abandon. In other words, the laws of physics themselves are violated by this ugly edge.

These regions of infinite curvature are grotesque enough that for decades people have argued that at the very least, if they exist at all, nature will censor them, hiding singularities behind impenetrable horizons like the event horizon of a black hole. I'm not the only one who recoils in the face of infinity. There are lots of anecdotes about the principles of cosmic censorship and infamous bets made among the leading thinkers with prizes of whisky and dirty magazines.[1] We all have vices.

General relativity points to its limitations and announces its failure at those regions of extreme curvature. A singularity also acts as a magnifying glass, amplifying small-scale quantum phenomena. Gravity, the

---

[1] Kip Thorne includes his bets as a subject topic in the index of his book, *Black Holes and Time Warps: Einstein's Outrageous Legacy*.

theory guiding the largest properties of the universe, is forced to confront quantum mechanics, the theory guiding the smallest properties of the universe. They don't mix well. Each resists the other, again in theory. The two most profound advances this century have not been fully reconciled. Like all great empires, one or both will eventually fall. Some new theory will emerge. How we live for that day.

What will survive the next revolution? There is no universal consensus on this and factions have broken off, each pursuing a refashioning of modern thought. I can't help but feel that geometry is the soul of gravity. Einstein's most brilliant insight was to suggest that spacetime is curved. He supplanted the Newtonian force with a theory of geometry: that is, a theory of curved spaces. The specific equations he eventually did derive may have only a limited validity, but I suspect gravity as geometry may survive even where general relativity does not.

This brings us everywhere at once chasing the most mundane and the most profound clues. It is not obvious how to proceed. Theorists, observers and experimentalists disperse like headless chickens, or maybe scattered ants, to do what they can, bringing their own strengths and

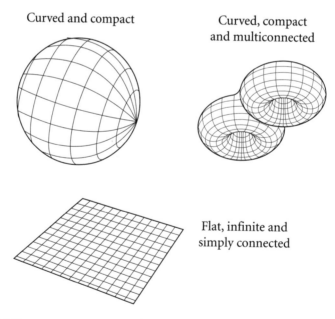

Curved and compact

Curved, compact
and multiconnected

Flat, infinite and
simply connected

Figure 9.1 *There are two aspects to the shape of space: curvature and topology. Topology describes global characteristics such as the connectedness of the space, the number of handles, and whether the space is compact or infinite.*

weaknesses together. Among the clues, mundane and profound, there are hints about the extent of the universe. There is another obvious and seemingly unrelated way in which relativity is incomplete. It's not even a complete theory of curved spaces. There are two facets to describe the shape of any space, including outer space: geometry and topology. Geometry describes the curvature of space. Topology describes the global shape and connectedness of space. The equations of general relativity tell you the local curves near any mass, but they tell you nothing about the global shape. These aren't the easiest concepts but they are the crux of this entire campaign so I'll dwell on them: geometry versus topology (Figure 9.1).

We have been considering infinite spaces, like an infinite, unbounded line. We can imagine a finite one-dimensional space easily enough. Take a piece of string. The thread is finite, but it is bounded. If an insect were to amble along the thread, it would come to an edge, namely the end of the string, and have to turn around. This is much like the fabled edge of the earth. What we really want is a one-dimensional space that is finite but *unbounded*. This we can also imagine in a loop of string. An insect can wander forever forward on this loop of string and never come to an edge, and yet the string is finite. The string is said to be topologically connected and compact.

The surface of the earth is a compact two-dimensional surface. If I left London and walked in as straight a line as I could, I would eventually come back to London (Figure 9.2). Maybe the universe is a three-dimensional version of the one-dimensional loop of string or the

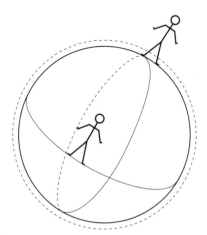

Figure 9.2 *Walking around the earth.*

two-dimensional surface of the earth. Maybe the universe is a three-dimensional topologically compact space. I have to admire so harmonious a resolution. Finite and edgeless, the cosmos would be elegantly self-contained. If I left the earth in a rocket and travelled in as straight a line as I could, I might eventually see the earth I left behind reappear on the horizon in front of me.

The mathematics of finite spaces is rather abstract but yields to remarkable visual descriptions. There is no limit to space that we can see, but maybe that's because we don't yet know how to look. Maybe it's just a ruse that the universe looks infinite, a trick of topological mirrors. Light wraps around the finite space and, in a small enough world, we could see our own images coming from all directions in the sky. Stranger than wonderland but not impossible. As we roll along I'll show you how the trick is done.

Topology may seem unrelated to the clash of the quantum and the gravitational, but it is not. We seem to stumble on these disconnected pieces but, as in a jigsaw puzzle, we only have to look a little to see where a given piece fits into the larger scheme. Quantum mechanics dismisses the idea of the continuum and with it in some sense infinity. Matter comes in individual quanta, finite and discrete bundles. Topology plays the same role for spaces. Topology describes the global shape of space, including compact, finite bundles of space. A quantization of general relativity will involve a quantization of space. Just as the continuous image of a light beam is recast under quantization in terms of individual lumps, photons, the continuous image of space is recast under quantization in terms of individual units of volume and area. It's sensible to guess then that the global topology of the universe must have been imposed at birth when the as yet unknown rules of quantum gravity were in effect.

Since general relativity only describes smooth, local changes in curvature, the topology of the universe will remain unchanged for the lifetime of the cosmos. The exception is in the form of singularities from the formation of black holes. But singularities signal the collapse of the realm of validity of general relativity. Over the regimes in which relativity is wholesome and true, the topology of the universe will be as it was at its quantum birth.

Einstein wasn't a mathematician in pursuit of the complete classification of spaces. He wanted a theory of gravity, and a theory of gravity he got. But if he was right, profoundly right, in discarding Newton's gravitational force in favour of a theory of geometry, we can finish the bible he started. We can write the missing passages, transcribe the history of the beginning, since it is in the beginning that topology would be fixed.

If we want to completely classify a space, we have to ask the mathematicians what they know and learn their language well enough to finish the cosmic story, document the beginning and prophesy the end. In their language, spaces are better known as manifolds. A manifold is specified by its dimension, its curvature and its topology. It's not just the curvature of the universe that is shaped by the laws of physics, not even just the topological connectedness or its size. Even the number of dimensions we occupy is an element in that still secret score.

Before we toy with the extent of space and challenge its mask of infinity, we have to wonder first about the dimensionality of space. Maybe the universe isn't even three-dimensional but just wears a very convincing guise. So the first property of the large-scale landscape of the cosmos we have to treat is the property of three: why three?

# 10

## ADVENTURES IN FLATLAND AND HYPERSPACE

Why three? Truth is, I don't know why. If it's any consolation, neither does anyone else. Being really honest, we don't even know for sure that three is the magic number. The universe is at least three-dimensional, but maybe there are more dimensions that we can't yet see.

So I brought you here on false pretences. I cannot tell you why space seems to have three dimensions, but I can tell you what dimensions are. We live in three dimensions so comfortably that I don't think it would be unfair to suggest that most people have never questioned what that really means. At least I can offer some idea of what it means to live in a manifold with different dimensions.

First, we can count dimensions constructively, going up the ladder of dimensions from zero to three. A zero-dimensional manifold is just a point and not really a space at all, in the sense that it has no extent whatsoever. Slide the point in one direction, say south to north, to make a one-dimensional line. Slide the line in a new direction, say east and west, as though laden with ink to make a two-dimensional square. Slide the square in the last remaining direction, up and down, to trace out a three-dimensional cube. While our visual perception limits us from continuing, we can recursively generate four-dimensional geometries by hypothetically dragging the three-dimensional cube into a hypothetical fourth dimension (Figure 10.1). Just as the line segment is bounded by two points, the square is bounded by four line segments and the cube is bounded by six square faces, the four-dimensional hypercube would be bounded by eight cubes. We can keep going and list the properties of generalizations of the hypercube in five dimensions, six dimensions or any number of dimensions. Mathematics can penetrate into realms where our eyes fail us and three dimensionality blinds us.

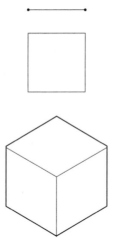

Figure 10.1 *Slide a point to make a line, a line to make a square, and a square to make a hypercube.*

We move around in three dimensions. There is a north and south, an east and west, an up and down. We can take a city map, which only depicts two dimensions, north–south and east–west, and stretch it flat on the floor. Most city maps tell you nothing about the up and the down, about the hills of San Francisco or the valleys of Los Angeles. Just because we can stand on a two-dimensional map, we still don't live in two dimensions. We can jump up and down on that map. We are three dimensional, no question. But what if?

What if we were only two dimensional – truly flat? Two-dimensional creatures are trapped on their two-dimensional manifold and can't jump off any more than we can jump off our three. A two-dimensional manifold then appears to be the universe. In principle, I can rewrite Einstein's equations in any number of dimensions, making them easier or harder to solve, and Einstein's equations tell us that all matter and all energy are forced to move along the natural curves in that space, whatever the dimensions may be. If we were two dimensional and lived on that flat map of the city, there would be no experience of the up and the down. No jumping off the map (Figure 10.2).

In a one-dimensional world, everything is completely linear. All creatures are line segments with a linear progression of one-dimensional organs. Their bodies are totally encased by their two endpoints. In a one-dimensional world, life is bleakly limited. Creatures can only move forwards, or backwards, and are forever stuck in alignment between the

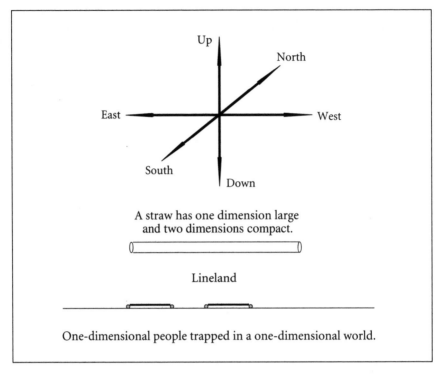

A straw has one dimension large
and two dimensions compact.

Lineland

One-dimensional people trapped in a one-dimensional world.

Figure 10.2 *Down the ladder of dimensions from three dimensions to one dimension.*

neighbours they were born with. Like infinitesimally thin beads on a string, all structure is forever trapped in order. A sealed house would amount to no more than two points, one in front of the inhabitant, one behind. A sealed house is an impenetrable prison: the one-dimensional creature could never leave, being completely unable to navigate around the door barring them from the rest of lineland. Even if they tried to break up the point particle door, the pieces would have nowhere to fly off to and would just form an equally confining one-dimensional wall of rubble. The only hope would be to vaporize the wall into a form of pure energy that the incarcerated creature could absorb without fatal injury. There's not much room for diversity in one dimension and you could argue persuasively that a one-dimensional universe simply wouldn't support life. Some degree of complexity is required for the development of structure and the building blocks of biology. It would be a pretty lame excuse for a cosmology.

Now up a dimension to two. Suppose there were a two-dimensional flatland where the indigenous population was totally unable to access the third dimension. Here I borrow from a truly peculiar book called *Flatland* by Edwin Abbott Abbott. It was written in the nineteenth century and chronicles a society of two-dimensional creatures. It would be impossible to discuss space, dimensionality and geometry without paying homage to this fantastical book. There is an odd tradition of writing books about this book. I prefer to read the source, not the abbreviated annotated notes on the source, but it would be insincere not to acknowledge and in some part recount the imagery of life in two dimensions composed by E. Abbott Abbott.

Flatlanders can move north–south or east–west, but there is no such thing as up and down. They can't point there and can't even find words to explain where the third dimension would reside. But from our privileged three-dimensional vantage point we can see that in a sense the third dimension is everywhere.

Flatlanders can only see each other edge on. Flatlanders are polygons: that is, many-sided closed lines. The polygon with the smallest number of sides is the triangle with only three. Then there is the square with four, the pentagon with five, the hexagon with six, and so on. A polygon with a huge number of sides looks like a circle. At least from the third dimension as we looked down on to Flatland we'd see a circle. To a Flatlander viewing their countrymen edge on, a circle looks like a simple line, but Abbott had a way for them to perceive perspective well enough to distinguish polygons. A Flatlander knew the difference between the receding arc of a circle and the straight edge of a hexagon. There's a sociopolitical take in the book where the more sides a person has, the higher is their social status. Abbott's many-layered fantasy satirized the inequity of his Victorian society. High priests were circles. Women were lethally low on sides, being mere lines. If a man came at them from the wrong angle, he could be mortally pierced by the imperceptible point of a woman's body.

Two-dimensional beings have more freedom to roam than their one-dimensional counterparts but will have some gastronomical peculiarities at the least. Any closed curve in two dimensions completely defines an inside and an outside. The body of a Flatlander is encased by a single bounding line. Any digestive track would cut the flat body into two separate halves, creating an obstacle to designing a two-dimensional creature whole and functional (Figure 10.3). In three dimensions we can sustain the digestive track with a cylindrical kind of topology and still remain connected and in one piece. It might not be hard to argue that the complexity of life, at least animal life, requires at least three

Figure 10.3 *The digestive tract of a two-dimensional animal.*

dimensions. We won't criticize their bodies, but we could invade them. We could use our third dimension to look into their innards. We could poke our extended fingers into their guts without piercing their skin. We would seem omniscient, able to transcend two dimensions and see right into the cross-sections of their flat houses, their sealed flat bottles, their flat bodies (Figure 10.4).

We could taunt the Flatlanders by appearing and disappearing in and out of their plane. We could pluck things from one room, making them disappear from Flatland and miraculously reappear in another room. We could take a left-handed glove, if only they had little hands, lift it off

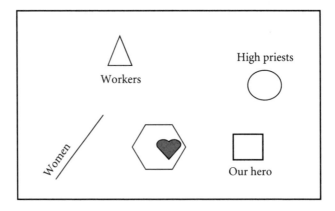

Figure 10.4 *Flatland (Edwin Abbott Abbott, 1884). All flat two-dimensional people are polygons, move and live in two dimensions, north–south and east–west, and are totally unaware of up and down. From the perspective of the third dimension, we could see right into their flat bodies and could poke their innards without piercing their skin.*

Flatland, flip it over, and drop it back down a right-handed glove. To the Flatlanders this would be complete magic. The glove would disappear and then impossibly reappear with a different handedness. We'd seem omnipresent and mystically powerful.

If we tried to reveal ourselves, they couldn't appreciate us in our full three-dimensional glory. They could only take in two dimensions at a time. If we tried to pass our hand into Flatland, they would see a series of appalling two-dimensional slices of blobs. If we stuck our fingers into the plane, they'd see five individual intersections. They wouldn't know that those fleshy cross-sections were connected at their roots to a unifying hand. Abbott prefers the Platonic solid to illustrate their experience of a three-dimensional invader. A Platonic solid such as a sphere dropped through Flatland first appears as a point as it touches Flatland (Figure 10.5). As the sphere continued to pass by, the flat creatures would only see the cross-section of the sphere and would be oblivious to its extension up and down. The cross-section would appear as a series of widening concentric circles. After the equator passed through the plane, Flatlanders would see the sphere as a series of shrinking concentric circles until only a point remained and then disappeared. Since high priests are circles, to all intents and purposes they would have witnessed the birth, growth, ageing and death of a high priest.

We could drop a collection of different solids through Flatland for sundry effect. A three-dimensional cube dropped through the plane would suddenly appear a square and then suddenly disappear. If we dropped the square from the vertex, a series of triangles would grow, mutate into a hexagon at the midpoint, then shrink and disappear. In this world of Platonic beings, we could change the gender of the apparitions in Flatland simply by picking the cube up, rotating it and dropping it back down at an angle, impressively demonstrating our influence.

In 1948 Orwell wrote *1984*, projected a hundred years to the future of

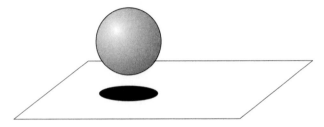

Figure 10.5 *A three-dimensional sphere intersecting Flatland would look like a solid circle. Fingers intersecting Flatland would appear disconnected.*

Abbott's 1884 publication. Maybe the Pythagoreans would find mystical significance in this association of numbers and anti-utopias. Orwell's Smith was freed only when his conviction was broken. Abbott's tale, like Orwell's, is a tragedy as the hero was imprisoned for heresy and lunacy, convicted on the basis of his insistence on the third dimension.


## I FEBRUARY 2000

Today England was at its most unsympathetic. Grey does not describe the weather. Grey implies a soft haze over the sky. Today the air itself is thick with grey, as though grey were a substance, a miserable emulsion suspended only in the dampest air. I retaliate by recklessly flinging my bike and myself down the canal path. Some days it's an obstacle course of snacking drivers, vengeful buses, mobile-phone-yapping pedestrians and roadworks. That's not even the dangerous part. The destination board at King's Cross was red with cancellations and faces were red with frenzy as we got the news of another derailment, another fifteen or twenty killed. The rest of us don't seem to know what else to do but board our trains. They throw up track numbers on destination monitors and change them with no warning. I'm amazed I've never been inadvertently carried off to Glasgow, but paranoia does me proud and keeps me ever on the alert, always ready for one of their tricks. They've got a nerve checking tickets today is all I can say. Despite all odds, I'm back in Cambridge and I'm alive.

Three-dimensional Cambridge. What if this isn't just fantasy fuelled by math? It is possible that Cambridge, London, the earth, the observable universe are just three-dimensional projections in a higher-dimensional space. Access to a fourth dimension could easily inspire fear and hysteria. A fourth dimension would be as mystifying and unimaginable to us as a third dimension would be to a denizen of two dimensions. If there is a fourth dimension, the fourth dimension is everywhere. A citizen of the fourth dimension would have seemingly supernatural powers and could poke into our insides without invasive surgery. They could access our brains and hearts and leave our skin unscarred. They could see into our three-dimensional houses, our sealed bottles, our bodies.

A flip in hyperspace could render any object into its mirror image. Your left shoe could be taken into four dimensions, flipped over and returned to you a right shoe. If someone could access the fourth dimension, they could take a left-handed glove, move into the fourth

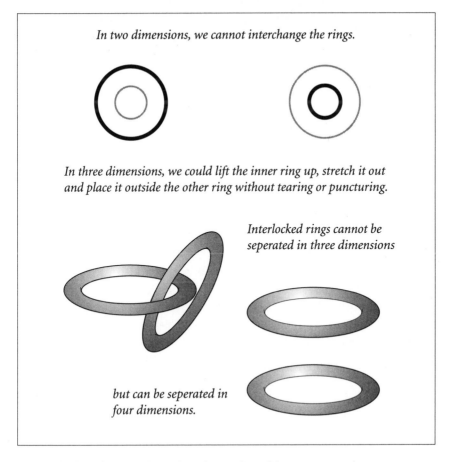

*In two dimensions, we cannot interchange the rings.*

*In three dimensions, we could lift the inner ring up, stretch it out and place it outside the other ring without tearing or puncturing.*

*Interlocked rings cannot be seperated in three dimensions*

*but can be seperated in four dimensions.*

Figure 10.6 *Topology can depend on the number of dimensions.*

dimension, flip it over and drop it back down on to three dimensions as a right-handed glove. They could take two interlocked steel rings, unhook them without the use of power tools and return the separated metal to us unharmed (Figure 10.6). Not just gloves and shoes but people too. Some twins are mirror images. One twin can be left-handed and the other right-handed with birthmarks on opposite sides. One of the pair could be lifted up into a fourth spatial dimension and returned truly identical to the other.

A four-dimensional hypersphere intersecting our three-dimensional world would look like a point initially, then a growing sphere, only to shrink and disappear again. We wouldn't know where to look to find out where it came from. Any manner of objects could be dropped from the fourth dimension, passing through our three-dimensional volume tilted and askew to reveal to our wide-eyed selves the miraculous appearance of smoothly mutating polyhedra that just as unexpectedly disappear.

### 3 FEBRUARY 2000

To a Flatlander, a three-dimensional object passing through Flatland looks like a two-dimensional object evolving in time. A circle being born, ageing and dying looks just like a sphere intersecting the plane of Flatland. This is how time is represented in the spacetime diagrams of relativity, which view spacetime as (3 + 1)-dimensions, three space and one time. In general relativity, different observers slice this spacetime diagram differently, which means they interpret a three-dimensional section as space and the remaining direction as time, while another observer in relative motion with respect to the first will interpret a different three-dimensional section as space and the new remaining direction as time.

Maybe there is a way to push the interpretation further and really view ourselves as four-dimensional objects that only perceive our inter-sections with a three-volume. So in a sense we are solid in four dimen-sions, there is no time, only the sequence of our perceptions as we intersect a three-surface that we interpret as the spatial world. It becomes clumsy trying to discard time. It's difficult to imagine motion of any kind without a concept of time. So even the sequence of intersections implies an ordering of that series which contains in it something we might as well call time. Any discussion of time has time ensnared in it.

Sometimes the smartest things are written in graffiti in toilets. I once heard a good anecdote about John Wheeler, the academic father to a generation of brilliant relativists including the particle physicist Richard Feynman. Story is he read in a public men's room the following prophetic words: 'Time is what keeps everything from happening at once.'

### 5 FEBRUARY 2000

If we stuck our fingers into Flatland, they'd appear as disconnected two-dimensional prongs. But we know those fingers are connected to a greater whole, our hands. If there were a fourth dimension, there could be something like that at work here. Individual three-dimensional blobs may be connected at their roots in four dimensions. To be extreme for the sake of argument, while we may feel fully encased in three dimensions and convinced of our individuality, our separateness, the isolation imposed by our skins, we may be like the fingers of a four-dimensional hand, connected and integrated at our roots in a fourth dimension.

A modern version of this scenario is conceivable where individual particles may really be endpoints of a string or a network that wiggles in the inaccessible extra dimensions. Maybe all things that appear distinct are really connected in hyperspace, like the tendrils of the aspens. A forest of aspens in Colorado are really one tree, their roots completely connected. It's like the *Monty Python* sketch where the crowd following the newly appointed Messiah, Brian, chant in unison after him, 'We are all individuals.' The one dissenter declares, 'I'm not.' Genius. Gives new meaning to the adage, 'I am one with nature.' Or as Woody Allen said, 'I am two with new nature.' Which is it, the one or the many?

From John Barrow I learned the games that could be played if only we had a fourth dimension. There were nineteenth-century spiritualists who claimed not only to access the fourth dimension but that they did so by communicating with a band of spirits that lived as material beings there. Those who wanted to believe their claims devised experiments for the mediums. The experiments involved turning a snail shell that wound to the left into a snail shell that wound to the right, interlocking two rings of different wood without severing the grain, and tying a knot in a ring cut from a pig's gut without breaking the ring. The popular American spiritualist Henry Slade failed to execute the assigned tasks but did conjure up other less impressive magical stunts. Not enough to free him from his gaol cell or pardon his conviction for fraud.

**7 FEBRUARY 2000**

I'm lying in bed unable to get up. Blankets cover half my face, but my arms are too comfortably hugging my sides like a living mummy so I don't bother to push the bedding away. I do peer above the stuffed duvet at the water-stained ceiling. I hear the repetitive riffs of a rehearsing musician, my rehearsing musician. We were gonna tell dad that Warren was a magician just so that he might actually be relieved to find out he's a musician. A magician like Henry Slade, a charlatan like Henry Slade. Even though Henry Slade couldn't access the fourth dimension, it doesn't mean a fourth dimension doesn't exist, or that we couldn't find it one day.

It's not impossible that there is a fourth dimension or a fifth, maybe even six extra dimensions or more. Extra dimensions recur in theories beyond Einstein's. Yet we seem to be oblivious of their existence. If they do exist, the burden that theoretical physicists face is to explain why we seem to only move around in three. The earliest suggestion was that these dimensions may be so small that we simply don't notice them; we have essentially no mobility in these compact directions, like an ant on a super-thin straw. If the straw is very thin in the compact directions, the space appears to only have extension in one dimension. By compact, I mean topologically compact, so there is a connection with hidden extra dimensions and the question we're asking about the topology of the large dimensions. The number and shape of all the dimensions are unknowns and inspire an active area of research in theoretical physics. Today we don't know, but tomorrow maybe we will.

Flatland was infinite, as I remember. If we panned away from Flatland it would go on forever, always filling our field of view. It doesn't have to be this way. If we look very close to the surface of the earth, it also looks pretty flat. But as we pull away into space and view the earth from afar, we see the earth's surface is a two-dimensional sphere. It is finite, compact and connected. Flatland can be finite and connected. So we have to ask: how different would Flatland look if it were finite? And the answer is very.

# 11

## TOPOLOGY: LINKS, LOCKS, LOOPS

It's not easy having a completely disenfranchised partner. Warren left school at fifteen. We tease each other that I'm going to have to home-school him, but in truth neither of us minds the disparity. There is no doubt the schism has widened here in England. I didn't foresee the times when I would be invited to high table in Cambridge on the same day he considers a job washing their dishes.

He knows nothing about Pythagoreans or geometry or infinity but has inherited a fixation on triangles and whole numbers. If all history were to be erased and a fresh civilization were to spring up without the burden of culture and the accumulation of centuries of knowledge, we'd inevitably rediscover triangles in the specific and geometry in general, irrationals and the hierarchy of infinity. It might just start with the rhythms of music or an obsessive compulsive tick of tracing polygons. But we do have that history and we can rattle on about Riemann, curved spaces and the dimensionality of manifolds.

The question of whether the universe is infinite or finite becomes a question of the universe's topology. General relativity cannot tell us whether the universe is infinite or whether it is globally connected, finite and edgeless. For this reason, the global shape of the universe has often been ignored by cosmologists and the universe was taken to be infinite. The faults with this assumption are numerous, and a recent wave of research has begun to take seriously the notion that the universe we live in is not infinite but finite.

Those of us interested in finite universe models had to become students of the mathematics of classifying the topology of spaces in order to make progress in cosmology. Mathematicians are like linguists and physicists are like writers. So the mathematicians invent these tools and show us how they work and then we drill and plumb with this arsenal of math to understand cosmic phenomena.

## 11 FEBRUARY 2000

Yesterday was a bad day. Couldn't sleep. I had a minor crisis at 2 a.m. wondering what I'm doing with my research. They're sending up satellites in a year or two which will scan the sky. From what they'll see they will be able to reconstruct enough about the history of the universe in a kind of cosmic archaeology to know if the universe is finite or if it is too big to determine. What if they see no evidence of the universe being compact? If the universe looks infinite as far as the eye can see, does that make my work useless? It still wouldn't mean the universe *is* infinite, just that we can't now, or ever, see to the edge. We'll never live long enough as a species. Of course, if we could see all the way around the cosmos, my knees would buckle. But even if the universe is not topologically finite, isn't there something still grand just in the thinking of it? Isn't it significant that pattern and symmetry and math and our eyes and our cosmos all fit together? Isn't that enough? I can hear the disembodied voices of some of my colleagues, a splattering of memories of so many conversations. I hear so many say the same thing with different words, accents and metre until the voices fall into unison and belt out: 'no'. They keep telling me it is not enough if you can't see it. At 2 a.m. I feared they were right.

I spend my waking hours trying to understand how number theory might transcribe some early universe pattern just because we can, and the fact that we can is a chilling fact. People sit at their desks scratching out calculations, always looking down at the paper, mining their own brain for the secrets of the universe. And there they are. They're in the structure of our logic like we've inherited the entire cosmic code. Thinking and calculating and thinking and calculating and then suddenly, there it is, understanding. Space stretches, time drags and, if only partially, we get an idea of some profound aspect of our universe. Scientists are always fond of saying that nearly all scientific discoveries are a response to experiment. But that's not really true. Maybe the details are anchored by experiment and this catapults us forward in terms of sorting out the details, but the big global scheme, the fact that math works and that we need it, rely on it to make sense of experiment or any other sensory experience, is the real marvel.

Mathematicians work for centuries trying to understand the topology of spaces and get Fields medals (a prize to honour great mathematicians and fill the gap left by the absence of a Nobel prize in mathematics) and do all kinds of obscure but brilliant work without ever caring if it has anything to do with the universe, with this one space, the biggest space of them all. Then a few cosmologists come along and realize we need these mathematicians and their ideas and maybe some of us get together and try to understand each other. That part is hard.

I was at this conference two years ago when topology started to become the main focus for a group of us. We all know each other, so it was no accident. When I was at the Canadian Institute for Theoretical Astrophysics in Toronto I used to organize these cosmology lunches. We would sit around a table and on one or two particular occasions we tried to make sense of some earlier work on topology with a researcher at Toronto, Igor Sokolov. At the time it seemed a tiring exercise and we dispersed without revelation. Still somehow within a year we were all working on cosmology and topology. A number of us had moved to new places – I was at Berkeley by then and Glenn Starkman and Neil Cornish were in Ohio, while Dick Bond and Dmitry Pogosyan stayed in Toronto. We broke up into three groups, more people joined up and a flurry of activity started. Which isn't to say that people were not working on topology before we did, it's just that we were all very conscious of each other's recent turn in thinking, so to us it felt new and in a way it was new, a new wave of ideas from a new community of people.

I for one was working on topology because these seemingly abstract spaces have a profound connection with chaos and it was chaos that I was interested in. John Barrow was visiting from England and both he and Joe Silk were intrigued about the cosmological implications and it just got started. Bond and I were on the phone one afternoon before the conference. We were lamenting how everyone's always accusing each other of stealing each other's ideas and when they're not doing that, they're stealing each other's ideas. And what's it all for, we were wondering. Maybe we were all wasting our energies. The possibility of a finite universe seemed painfully remote, but as soon as the words were out my mouth I realized I was wrong. It was one of those rare moments when I saw my belief system shift. I could not justify a belief in an infinite universe created from nothing. The universe must be finite. And though this was maybe no more than a belief fuelled by a raw and shared instinct, I was shaken. By the time I got to the conference I was in a bit of a state, but so apparently was everyone else.

The meeting was small. Maybe twenty-five or thirty people. Half were cosmologists and half mathematicians. To the astronomers a compact universe is complete exotica, but to the mathematicians it is practical application. The participants were acting as though they had five minutes to live and they had solved Fermat's last theorem. People were feverishly turning to whoever was closest and frantically outpouring what they knew and they were begging for help with what they didn't know of manifolds and symmetry groups and differential equations and what all that had to do with the universe. You'd think we had to save the world and the clock was ticking. There was a collective hysteria.

It was a rare display of genuine curiosity and I attribute that to the mathematicians. From where I'm standing their community seems much less greedy and more cultivating of ideas. I haven't seen a physicist publicly acknowledge that they didn't know something in an age. Arms were grabbed to get each other's attention and people would pry their faces away from one critical conversation to latch on to another. We stood stupidly packed together, occupying a tiny space in a big room, pained by every discussion missed.

That day was a good day.


### 13 FEBRUARY 2000

The academic life is so good sometimes it seems ridiculous to complain. Then other times I can imagine the dust collecting in sandpiles on their tweed shoulders as they sit ever-so-still in the most inhumane buildings ever constructed, trying to convince their colleagues of their worthiness. Promotions in this field can feel like a forced retirement to a dusty old remote corner. It panics me as I'm warned of the recklessness of resistance.

I don't know when I started clutching notebooks everywhere I went. Maybe it was a form of resistance. I noticed that it had become chronic some time ago when I sat with Rory over fried eggs in Los Angeles and felt compelled to write down everything we were saying. I met Rory Kelly on my way to the Aspen Center for Physics. He was on his way there too with Neil Jimenez, which is pretty unlikely since they're both writers and directors. They were on a reconnaissance mission to gather information about cosmologists for a potential movie script. I don't think the cosmologists provided much raunch or suspense or fodder for a script. I tried to compensate with salacious rumours that they efficiently ignored. We've been tumultuous but great friends ever since and I am ever so adoring of Rory although he badgers me and often gives me a terrible time. As he did that morning over the fried eggs in LA.

Rory tries to explain to me a theory of humour which is based on cruelty, and he can't believe I don't know anything, not even the most elementary theory of comedy. So he explains about Groucho Marx's elephant. The key ingredient in the joke is the aggression. I shot an elephant in my pyjamas. How he got there, I don't know. All comedy is at somebody's expense. In this case, the elephant's. I think this is his way of apologizing for this morning's comedy being at my expense.

I write this down in my notebook.

Rory reminds me that I was nothing but an astrophysicist when he met me and I remind him that I'm still nothing but an astrophysicist. And because of this I didn't know about Groucho Marx's elephant. What I know is how to ask questions. At the time I was uncomfortable with questions about topology and its application to cosmology. So he makes me explain my questions.

The topology of the universe is the hardest topic for me to approach in this way. I'm too close to it. My introduction to topology was all symbols: $M = G/\Gamma$, $\gamma \in \Gamma$. Of course this is meaningless to most. It was meaningless to me once too. Then when I finally understood this language, I learned to hang English words on it. The translation into words is more manageable: geometry is all about curves and smooth changes in curvature. Topology is any aspect of the shape that does not change when the curvature is varied.

The simplest way to demonstrate the distinction between topology and geometry in these definitions is also the most deceptive. I think it is simplest to get a feel for topology by using examples of two-dimensional surfaces. Two-dimensional examples appeal to our tactile senses. We can hold these in our hands and see them with our eyes. I'll use them first to lend an intuition, but then we'll have to step up a dimension to three-dimensional surfaces, which we can't hold with our hands or even visualize in full, not even with our adept imaginations.

Start with an egg. An egg is actually a three-dimensional object, but ignore the inside completely. Focus instead only on the shell. The shell forms a two-dimensional boundary to the egg. This two-dimensional surface is curved. If the shell were made of clay, I could smoothly change the curvature to make the shell perfectly round or oblong. But there is one feature I couldn't change by smoothly rolling the shell in the palms of my hands. I couldn't change the fact that the shell is compact. The shell is completely closed and connected to itself. This is a topological feature. To change the topology of the egg, I would have to break it. Breaking, tearing, gluing, these actions change the topology.

Now take a ring. The surface of a ring is also finite and connected. I could stretch a ring made of clay to fit a fat finger or shrink it to fit a skinny finger. I can smoothly change the curvature of the ring to make the two-dimensional surface round or alternatively square. But when I change the curvature of the surface, one thing that won't change about the surface of the ring is that it is compact and connected. The connectedness of the surface is a topological feature.

Even though both the ring and the egg shell are finite and connected, they do not have the same topology. The ring surrounds a hole and the shell does not. To change the shell into the ring would require puncturing the shell (Figure 11.1). This act of tearing is not a smooth change in curvature and constitutes a change of topology. To distinguish a doughnut hole from a cut in space, these kinds of holes are called handles. The number of handles or holes is a topological feature called the genus of a space. The shell has genus zero. The ring has genus one. A pair of eyeglass frames has genus two.

The topologist would not know the difference between a doughnut and a coffee mug. Since they are both compact and both have one handle, they both have genus one, despite their very different curves. I can smoothly mush a coffee cup made out of clay into a doughnut (Figure

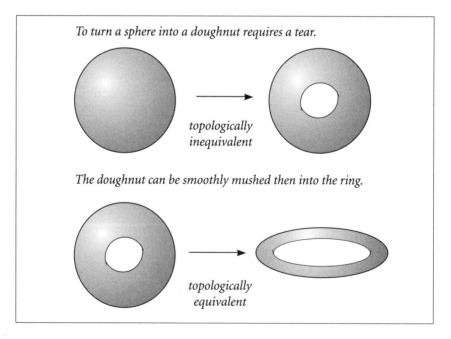

*To turn a sphere into a doughnut requires a tear.*

*topologically inequivalent*

*The doughnut can be smoothly mushed then into the ring.*

*topologically equivalent*

Figure 11.1 *Equivalence classes.*

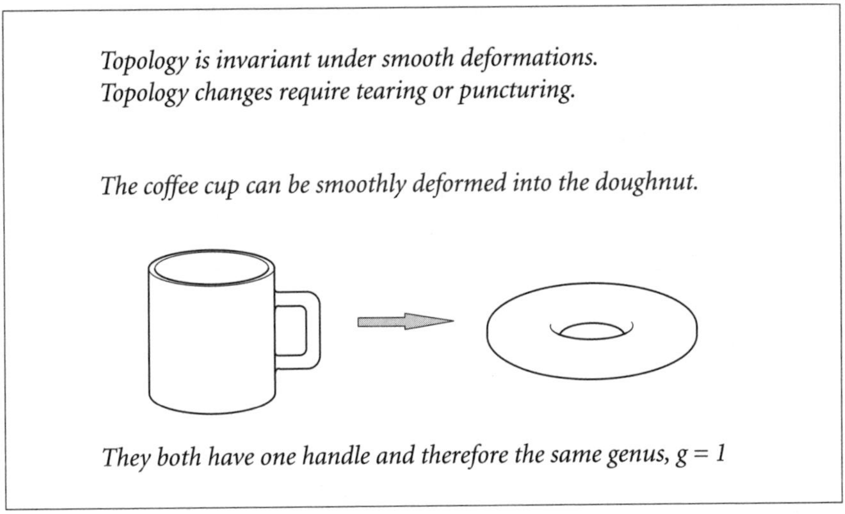

*Topology is invariant under smooth deformations.*
*Topology changes require tearing or puncturing.*

*The coffee cup can be smoothly deformed into the doughnut.*

*They both have one handle and therefore the same genus, g = 1*

Figure 11.2 *A coffee cup and a doughnut have the same topology.*

11.2). To make the clay doughnut into an eyeglass frame, you would have to puncture another handle in it. This is not a smooth deformation, and it amounts to changing the topology. So while doughnuts and coffee mugs are topologically equivalent, eyeglasses and doughnuts are not.

There is something deceptive about these examples. We manipulate these two-dimensional boundaries in three dimensions. We can look at them from our three-dimensional vantage point, roll them around, see with our eyes that the ring traces out a handle. This is an artefact of our physical three-dimensionality. What if we only had two-dimensions and there did not exist a third dimension? A two-dimensional surface can still exist, still be finite and still wrap around a handle. Only now it becomes more difficult to imagine and more difficult to visualize. Difficult, but not impossible. This is our challenge: how do we classify compact two-dimensional surfaces without bending them into our additional dimension? How will we know they have handles or holes?

This is critical for a study of the shape of the cosmos. We cannot jump off our three dimensions. We cannot manipulate three-dimensional spaces as though they were bent into four dimensions. We don't have four dimensions at our disposal, at least that we know of, and we can't just step away from three dimensions to look and see handles. We have to rise up to the demands of abstraction and to get there I'll go back down the ladder of dimensions, starting with one.

A one-dimensional manifold is like a line. A line can be curved like a wiggled piece of string or straight. In principle it could go on forever or it could be compact like a circle. I can stretch the circle into an oval and I've changed the curvature but not the topology. This self-connected loop of string, bounded and finite, is the only one-dimensional topology possible besides the unbounded infinite string.

The circle is finite. It has no edge. There is a sense in which it creates a handle when we view it drawn on a two-dimensional sheet of paper. A one-dimensional creature living on the line, walking around and around, is not privy to view this handle, but she can still perform experiments to determine the topological connectedness. She could tie a string around her space and she would realize there was no way to shrink the string. It would catch around the circle and stand firm (Figure 11.3).

Light is as bound to space as any matter. A one-dimensional creature alone on a one-dimensional compact loop looking forward would see light from her back and looking back would see light from her front, in as much as she had a front and back. She'd lay there trapped in this

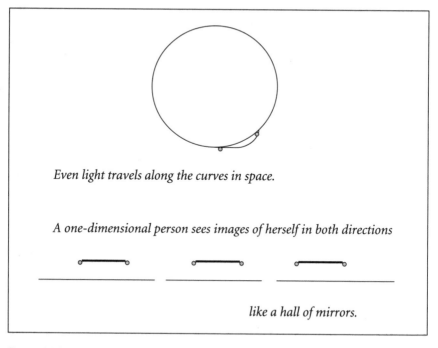

*Even light travels along the curves in space.*

*A one-dimensional person sees images of herself in both directions*

*like a hall of mirrors.*

Figure 11.3 *A compact one-dimensional space has the topology of a circle. An equivalent representation of the flat one-dimensional space is an infinite line of tiles. Each tile is a ghost copy of the original.*

world of limited possibilities looking at herself for eternity. If she didn't know she was alone on the loop, she'd soon find out. Having never seen herself, she might mistake her own image as the image of another. But if she moved towards the image, she'd see the image move away. If she tried backing up towards the image she sees behind her, she'd see the image move away. Instead of saving her from isolation, the reflection would retreat. If she marked her place of rest with a red dot and pursued the ghost she saw in space, she would find herself racing around the loop and returning to the red dot without ever colliding with her own shadowy image. Her solitude would be total.

The circle topology does not require three dimensions or even two dimensions. Imagine not that the flat line is a curved circle living in two dimensions, but rather that it is space itself. Take a straight line segment and identify the edges. By identify the edges I mean that the leftmost point is really physically the same as the rightmost point. A one-dimensional creature moving left at the leftmost edge would just re-enter space from the right. The edge does not really exist. Her transition is smooth and imperceptible. This identified line has the same topology as the circle but has no curvature; it is truly straight. It does not live in three dimensions. It is space itself.

The circle has mythological significance as a symbol of eternity, sometimes manifest as a serpent holding its tail in its mouth. There is an element of the infinite in this finite space. The universe would look deceptively infinite. The best way to visualize a one-dimensional straight loop is by tiling (Figure 11.3). A tile in this context is a straight line segment. If the straight line is topologically compact, then the left edge is glued to the right edge, without bending the segment. This is completely equivalent to gluing an identical copy of the tile, left edge to right edge, and another identical copy of the tile, right edge to left edge, and so on. The infinite length of this one dimension can be filled with an infinite number of tiles glued edge to edge in both directions. This is just a method to visualize the compact space with the topology of a circle without artificially bending it into two dimensions. These tiles are not new spaces but are truly identical copies. The solitary one-dimensional dweller in the centre of the original tile would see a copy of herself in the centre of every other. If she moved left, so would all of the copies. If she moved right, so would all of the copies. If she tried to beckon to the others, the infinite images surrounding her would beckon the same. A looking glass stranger than Alice's, her finite world looks like an infinite series of replicas. They are just chimera, ghost images.

The further away the tile, the longer it would take light to reach the imprisoned observer. The most distant images she could see would have come from the longest time ago. She could watch herself evolve, literally watch her life pass before her eyes, whatever limited process of development a one-dimensional creature can endure. But despite the infinite parade of images she would see, she could know that the universe was finite. Marking her home with the red dot and exploring what seemed to her a vast and populated line, she could travel through her tile, into the neighbour, only to find her red dotted home at the end of her search and herself still the sole occupant.

## 17 FEBRUARY 2000

Moscow. I'm in a sanatorium. This is the craziest place I've ever been. It was all going so well. I got through passport control in Moscow. The conference organizers sent a student to the airport who knew enough English to get me through customs. His nice mom drove while he guided me through the centre of Moscow and then right out through the other side of the city up an oddly wooded road to a chained gate. He kept saying the conference was in a sanatorium, which seemed a fair enough linguistic mistake albeit an unforgivable mistake in conference planning. But there's no question, the conference is being held in a sanatorium. Rolls of carpet worn thin unravel down long halls with shutters for doors. An old woman sways at the entrance, holding a cat. If I were to try to make a place look like a sanatorium, I'd hire a swaying old woman holding a cat. I'm looking for Eherenfest and Boltzmann here.

My room is frightening and huge and cold. There is a portable heater that I park precariously close to the bed. Is an incendiary end better than hypothermia? I weigh the odds. There are stains of dripped blood on some of the bedding. I take covers from one bed and sheets from another until I get a set without such a conspicuous memory of the previous inhabitants. I've been to dozens of conferences around the world. Sometimes I stay in four-star hotels, sometimes in dorm rooms, but this is my first mental ward. What's that saying? 'I've been rich and I've been poor. Rich is better.' Am I just a capitalist? There is real beauty here, which can be pieced together in the broken tiles, the tattered yarn, the classical paintings. There was opulence here once. I see the Western Europeans scoff at the cold plates of food and disparage our unusual accommodations and the exorbitant conference fee demanded only from the foreign participants. Our sense of entitlement embarrasses me. We pay in US dollars, although I think I'm the only American.

These sweet old women wearing babushkas bring us our dinners and their eyes gleam with amusement and tenderness. They speak to us in Russian and we reply in English. Even though there are only a couple of native English speakers, it's our only common language. Conversations break out in several languages but in the end it's all Greek anyway. Literally Greek: $\mu$, $\tau$, $\alpha$, $\beta$, our common language in Greek symbols favoured by mathematicians. I need to be alone. I'm having a mild anxiety attack. We're so pampered. So greedy. So indulged.

Sometimes it's good to be physically tortured. This weird isolation with no distracting entertainment is a challenge. I have run out of books. When I reached the back cover of my last novel, I flipped it over and started again. There are no newspapers in my room to distract me, no food, no phone, no booze. Just my person in a cold room. I'm even running out of paper and my pens have actually run dry.

Now, Eddie Murphy is on the TV. The way they dub films in Russia is to shout over the original language. It takes me a while, but I realize it's the same guy doing all the voices. I can't get out. The gates are locked along the perimeter of the grounds and even if I were to scale the metal fence, I'd be in the woods. I'm gonna watch Eddie Murphy and listen to the Russian guy and then go to sleep.

## 17 FEBRUARY 2000

I can't sleep. I will lie here patiently and think. There are people who argue that there's a way physicists think. I'm always offended by those arguments. Who are they trying to convince and why? Is there a litmus test for a career choice? I don't care one way or the other, but I do admit that we can watch each other and learn, not just the facts but the methods, and I confess that the methods I admire most hinge on the knack for simplification. We are trying to understand the most confounding, complicated conglomerate – the universe. That's too hard, at least at first, so we talk instead about points and lines, circles and one dimension. We teach ourselves how to take the first unsure steps on unfamiliar terrain. We get sturdy and more confident and brave the next challenge.

Before we describe the possible shape of our own universe, we still have lessons to learn and skills to acquire by simplifying space to two dimensions. One-dimensional space is limited. Two dimensions are more flexible. The flexibility provided by the additional dimension

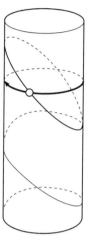

Figure 11.4 *Walking around a cylinder.*

allows for an infinite number of possible topologies in two dimensions and so the task is to classify them and find a way to count them.

In zero dimensions there was the point. In one dimension there was the infinite line and the compact circle. In two dimensions there are topologically distinct spaces that are simple to describe and can be used to generate the remaining infinite number of spaces. The list begins with the infinite plane, the infinite cylinder and the completely compact torus, which looks like a doughnut. A creature living on the cylinder could travel forever along the length of the surface but would discover the one compact dimension by walking straight around the circumference, only to return to where she started (Figure 11.4). A resident of the torus could detect the finite extent of his surface by travelling around the space in as straight a line as possible. He would take a path that loops

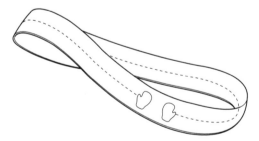

Figure 11.5 *A left-handed glove taken for a trip around the Möbius strip can be turned into a right-handed glove.*

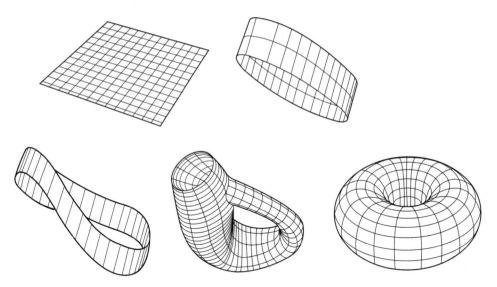

Figure 11.6 A list of two-dimensional topologies: the infinite flat surface, the infinitely long cylinder, the infinitely long Möbius strip, the compact Klein bottle and the compact torus.

around the space – a road that eventually leads home. There's a limit to how far away he could ever go.

There is also the Möbius strip and then there's the Klein bottle. The Möbius strip is non-orientable, which means that if you started walking on the outside of the strip, you'd end up on the inside after one full circuit and only after two full circuits would you end up on the outside again. The Möbius strip has no real outside or inside as both sides constitute the continuous surface.

If the strip is completely transparent, a left-handed glove taken around the space becomes a right-handed glove (Figure 11.5). This inability to permanently orient left from right is what earns this space the description of non-orientability. I imagine the terrible twins running around a Möbius strip, turning left-handed into right-handed with each circuit around the space. The Klein bottle has one handle like the torus but is non-orientable, like the multiconnected but infinite Möbius strip (Figure 11.6).

Then there's the sphere (Figure 11.7). The sphere is simple in the way it manages to consolidate its shape. It has no handles or holes. The inhabitant of the sphere would know there were no handles or holes by tying a string around the circumference of space. He could shrink this

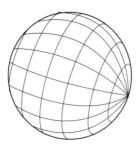

Figure 11.7 *The compact, finite spherical surface.*

Figure 11.8 *A rubber band around the sphere can be slid to a pole and contracted to a point. There are no handles to catch onto.*

string down to a point by sliding the string to one of the poles without catching it on any holes. So the sphere, though quite fully connected, is said to be simply connected (Figure 11.8).

A two-dimensional creature living on the sphere would experience life like on the earth. If she walked in a straight line, she'd come back to where she started. Only it's more dramatic than the earth because light is not forced to follow the curves on the earth. The earth is a roundish rock that lives in three spatial dimensions. If we shine a torch, the light will move in three dimensions off the surface of the earth. By contrast, in general relativity, light is forced to travel along the curves in space. If the

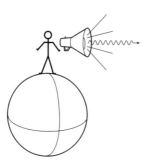

*Light does not travel along the curves of the earth's surface but rather moves freely into space.*

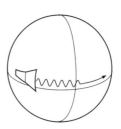

*By contrast light is forced to travel along the curves of a two-dimensional compact space.*

Figure 11.9 *Shining a torch off the earth versus on the surface of a two-dimensional spherical space.*

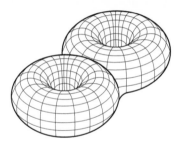

Figure 11.10 *Two connected tori make a double-doughnut.*

sphere is a universe, then light will wrap along the great arcs of the sphere. Shining a beam of light along the surface of the sphere, the light will follow the great curves and eventually return near its place of origin (Figure 11.9).[1]

The spaces of Figures 11.6 and 11.7 can spawn an infinite number of two-dimensional spaces. Two connected tori make a double doughnut with two handles (Figure 11.10). A string of tori make an infinite number of topologies with an ever-growing number of handles. A connection of tori or spheres together with the Klein bottles, etc. generates another infinite list of topologies.

The topology of all of these spaces can be tested with rubber bands or contractible loops of string. All loops of string can be shrunk to a point on the infinite plane and on the sphere, and so they are said to be simply connected. On the infinite cylinder, a string could catch around the circumference and therefore the cylinder is said to be multiply connected. The string cannot be removed without cutting the thread, so this property of loops and their reducibility probes the topological character of the space. In addition to the loops encompassing the diameter, the torus can find a second irreducible loop. A string threaded through the hole of the torus will detect the handle (Figure 11.11). No matter how much the string stretches or deforms, it will wrap around the hole once. It cannot be undone without cutting it. The larger the number of handles, the larger the number of such distinct loops. Even though the two-dimensional inhabitants can never see their space from the outside, they can in principle detect the topology of their universe with this search for incontractible loops.

Meanwhile, it's morning back at the sanatorium. I've never worn a

---

[1] Another peculiar space called the projective plane can be made by identifying opposite points on the sphere.

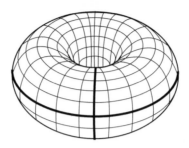

Figure 11.11 *Irreducible loops can detect the handle of the compact torus.*

watch. The ticking drives me out of my head. And even if I wear a silent clock my eyes fixate on the inevitable passage of time, second after second, accumulated minutes until I have to wrench the wretched thing from my wrist. So here I am, in Russia, no clock, no watch, no phone. I don't understand the television. I can't communicate with the staff. I have no idea what time it is. I try to resurrect my astronomical knowledge to determine the hour from the minimal natural light. But I'm a theorist and I still have no idea what time it is. I make a guess that it's time for the conference day to start and go down to the meeting room. It was a poor guess. It was only 6 a.m. I wait.

I'm here to celebrate Isaac Markovitch Khalatnikov's birthday. It was Khalatnikov with Lifshitz and Belinsky who suggested that the big bang would begin chaotically. Khalatnikov is so lovely and makes me an honorary member of the Landau family and tells me about their hardships, and tells me how when he loves somebody he really loves them forever so I should be cautiously pleased. I am abundantly pleased. I have admired his work and history and am delighted to meet the real man. After a while I'm feeling better. The sanatorium is working and I feel my mental health recovering. The people here seem warm and earnest, serious and sincere. They keep all their richness on the inside.

# 12

## THROUGH THE LOOKING GLASS

**18 FEBRUARY 2000**

Warren's father was murdered. He used to tell the murdered-father story while on stage, stirring nervous shocked laughter in both the audience and himself. A deep memory of being ensconced in warmth and home and family is tainted for him. It's simpler for me. California has spoiled me and I've grown to hate the winter chill, but I remember how nothing is sweeter than coming in from the cold. I remember warm car rides home and the serenity of dad driving as I watched the street lights or dozed in the back.

Warren and I were in California when he started looking for the father he misplaced nearly thirty years ago. He was braced over the wooden balcony of our San Francisco apartment, watching me drag my bag of books up the stairs hampered by a handful of crumpled scratch paper that I sloppily filled with notes during the rush-hour commute. The news from England had arrived. 'Dad's dead', he told me when I made it to the top. When the shock began to fade we took drives down the California coast to fend off despair. The following weeks were filled with lost English uncles, unknown Irish grandparents, family trees and stories of disappointment. By some odd series of small steps, Warren's back in England now, wondering what he's doing there and where he belongs. I'm in Moscow, wondering what I'm doing here and where I belong. England was the scene of his history and Russia the scene of mine. We've both lost our distance and with it our perspective.

**19 FEBRUARY 2000**

Distance and perspective are everything. With distance and perspective I can understand life on a connected surface fairly effortlessly, while the same visual is incomprehensible to a two-dimensional animal trapped on the surface, its vision blurred by sheer proximity. We can hold all of these two-dimensional surfaces in our hands, roll them around, hold them at arm's length. All of these two-dimensional surfaces are easy to visualize because we can view them from the outside. Both distance and perspective are lost when we try to view our own universe. We can't jump off space and see it from 'outside'. There is no outside. We will move along the curves like a travelling stain migrating around the fabric of space. How can we begin to imagine a connected three-dimensional volume? The most important aid I know of is the method of tilings, and so before we face three dimensions we'll learn how to tile two dimensions.

The sphere, the torus and all the other two-dimensional topologies appear to bend into one higher dimension, in this case three. But math, and luckily mathematicians, are not restricted by this predilection of our faculties. These are truly two-dimensional objects and we don't have to draw them as though they protrude into three dimensions. The machinery for visualizing and describing these multiconnected topologies in two dimensions alone, without any reference or need for a third, is absolutely essential for progressing on to three-dimensional topologies.

Specifically, we can make two-dimensional Flatland compact while preserving both its flatness and its two dimensionality. The best place to start is by making a curved doughnut and then seeing where we erred. The do-it-yourself instructions for building a compact space have three stages. In brief, the instructions say to start with a certain geometry[1], say flat space. Cut a shape from that cloth, say a rectangle. Then glue the edges of that fundamental shape together. Glue the left edge to the right edge to make a cylinder. If the cylinder were infinitely long, it would represent a multiconnected but still infinite surface. Then glue the top to the bottom of the rectangle, to make a torus that is topologically identical to the doughnut (Figure 12.1). It is compact and multiconnected. The seam disappears on the surface and is completely meaningless.

---

[1] Here a distinction has to be clarified between geometry, which describes the curvature of a surface, and topology, which describes the way that shape connects to itself. There are three geometries of constant curvature in two dimensions: flat space, positively curved space and negatively curved space. Now we want to find spaces that preserve these smooth two-dimensional geometries but have the same topologies as the bent surfaces in Figures 11.6, 11.7 and 11.10.

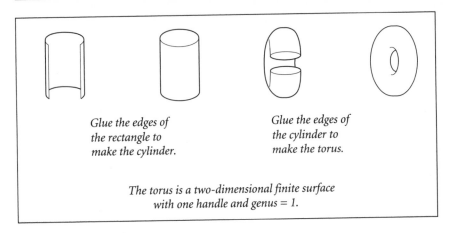

Glue the edges of
the rectangle to
make the cylinder.

Glue the edges of
the cylinder to
make the torus.

The torus is a two-dimensional finite surface
with one handle and genus = 1.

Figure 12.1 *Gluing the square to make the torus.*

When we started with a flat rectangle to build the torus, we made a compact space but it wasn't flat. What we really built out of the rectangle is a torus of revolution. Where did we err? We bent the sheet of paper to visualize nesting the two-dimensional surface in three dimensions. But we don't have to bend it into three dimensions, and this is where it gets clever.

We could have made a torus out of the rectangle by mathematically imposing the rule that as a path leaves the left edge it must re-enter the right and as it leaves the top it must re-enter the bottom. This is how many video games work. Video animations fly off one edge of the screen only to re-enter from the opposite edge (Figure 12.2). The implicit rules for interpreting a map of the world also work this way. The earth is a compact, round surface bent into three dimensions. We can project the globe onto a flat map with the rule that the edges of the map aren't real – there are no seams on the earth – but instead are identified. When we read a map of the earth we can pan west from California across the Pacific and eventually reach the west-most edge of the map with the understanding that we are to re-enter from the east-most edge of the map.

Similarly, a flat two-dimensional map of the torus is a rectangle with identified edges. The edges of the identified rectangle are no more per-ceptible to an inhabitant of the flat two-dimensional torus than they are to the inhabitant of the earth. If a guy stuck his arm out the right edge, he would see it re-enter around the left edge, but he would experience no physical sensation at passing through the line, since the left edge is truly identified: that is, identical to the right edge. Further the two-

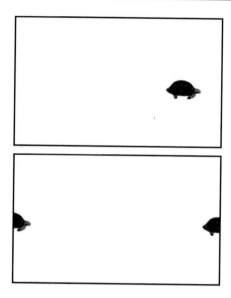

Figure 12.2 *Life on the flat torus (reminiscent of* Flatland, *1884, by Edwin Abbott Abbott).*

dimensional surface is truly flat, there is no curving of the surface and it does not require the third dimension to exist at all (Figure 12.3).

There is still a difference between the flat map of the earth and the flat map of the torus. The earth is just a round rock and not a curved space in the sense that Einstein invoked in general relativity. All matter, all light and all energy will follow the curves of the universe. If the torus is to represent a universe, then the flat rectangular map will show the path of light rays through space. A reflection of Flatland will travel across the finite space, exit one edge and re-enter the other. So a Flatlander would see a ghost image of himself – many ghost images of himself.

An even better way to visualize what a Flatlander would see if Flatland was compact is to tile space just as we tiled one dimension to represent the compact circle. To tile two dimensions, begin with the flat rectangular map of the torus. In the map, the right edge of the rectangle is identical to the left edge. We can enforce this rule by taking an identical copy of the rectangle and gluing the right edge to the left edge of the original tile. Continue to glue copies until the entire plane is filled without gaps or overlaps. Akin to tiling a flat wall or a flat floor, rectangular tiles cover the two-dimensional space. Unlike tiling a bathroom, these tiles are not just similar but are truly identical copies of Flatland. If you stand in the middle of the original tile, you will see yourself standing

*By embedding the two-dimensional torus in three dimensions we had to bend it and so curve it*

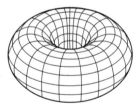

*whereas the flat identified square lives in two dimensions and is everywhere flat.*

*They have different curvatures but the same topology.*

Figure 12.3 *Curvature versus topology.*

in the middle of your neighbouring tile (Figure 12.4).

Tiling offers another true representation of the torus in the sense that it does not distort the geometry with any false curves. If you move to the left, you will see yourself move to the left everywhere in space, although the finite light travel time will delay images from more distant tiles.

The Flatlander could deduce that the world was flat and compact and even determine its size. Since light itself has to move along the curves defined by the space, it will try to travel in a straight line on the torus, exiting one edge and re-entering the other according to the identification rules. If an observer looked forward she would see the back of her head and looking up she'd see her feet. She would have the illusion of seeing copies of herself in every direction (Figure 12.4). The more an image winds around the space, the further away it appears, so she would see a collage of images of herself at different ages. If she ran towards her own image, that image would race away from her and in a maddening and unwinnable game of tag she would chase her own reflection through the confounding hall of mirrors.

This illusion of distance gives the inhabitant of a compact space the impression that her universe is actually infinite and not finite. If she

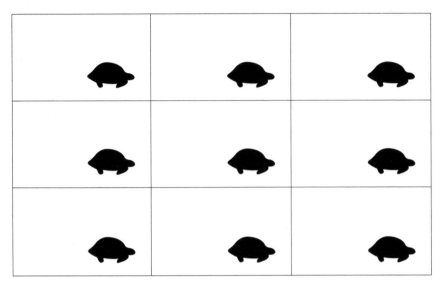

*Like a hall of mirrors, the turtle sees identical copies of himself in all directions.*

Figure 12.4 *Tiling flat space with the rectangle: an equivalent way to visualize the compact torus is to tile the flat plane with identical copies of the space.*

looked back far enough, she could watch her childhood replaying in the distance. She could see her life unfold and the lives of her ancestors.

We can make any of the topologies of Figure 11.6 out of a flat rectangle. We could have glued the left edge to the right edge after rotating the sheet by 180°, to make a Möbius strip. Gluing the top and bottom then creates a fully compact Klein bottle. Bending the Klein bottle into three dimensions creates an odd-looking curved surface that has to self-intersect (as in Figure 11.6). But like the flat torus, there is a truly flat Klein bottle, content to live in two dimensions. The bottle is neither curved nor self-intersecting.

The truly flat Klein bottle can also be represented as a tiling of the flat plane with each tile rotated by 180° in one direction to create the reflected and translated pattern of images specific to the non-orientable bottle (Figure 12.5). The distribution of ghost images is different in the Klein bottle from in the torus. Due to the twist in one direction, every other ghost image in that rotated direction is upside-down, and so the pattern of clones is a clue to the geometry and topology of the universe.

The tiling picture reveals that a finite space looks like a hall of mirrors and the kaleidoscope of images can reveal how the tiles fit together. The

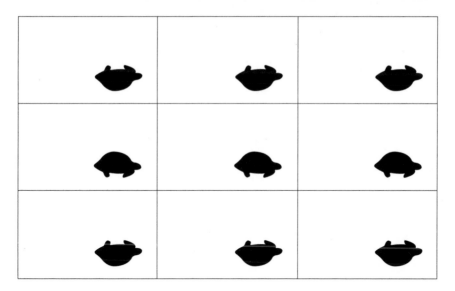

Figure 12.5 *A tiling representation of the Klein bottle.*

further away the tile, the longer it takes light to get to observers and so the older the image they'd see. They wouldn't have to rely on memory or the stories of history to place themselves in context. They could use binoculars or telescopes to look back and watch that history play itself out. History would become an observational science, a branch of astronomy. If we wanted to know who shot JFK, we could figure out how long it would take the light from 1963 to cross the finite space, build a strong enough telescope, point it in the right direction and watch the ghost images replay that horrible day.

Viewing compact surfaces from 'outside', from the distance of three dimensions, offers us privileged insight. We have an instant picture of the way the spaces connect to themselves and their array of handles. But if we gain something, we lose something too. Going back down into two dimensions and peering at the world without distance and without the aid of a third dimension, exposes a finite universe to be a perverse looking glass.

## 21 FEBRUARY 2000
I'm still in Moscow[2]. The tiles in the bathroom distract me. I'm in the

---

[2] Honestly, I've gotten the dates wrong. The conference in honour of Isaac Markovitch Khalatnikov was in October 1999.

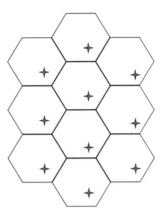

Figure 12.6 *Tiling the flat plane with hexagons. The plane can only be tiled with hexagons or parallelograms. A torus can also be made by gluing opposite sides of the hexagon.*

tub and the water drips audibly from the faucet like a lazy rhythm section. The tiles wrap inside of the bath and out. They are squares, shaped like the tiles we used to tessellate flat two-dimensional space. The floor is covered by hexagonal shaped tiles. They fit together smoothly and without gaps. When I first started working on topology, I saw hexagons everywhere – hexagons in San Francisco train stations and old-fashioned bathrooms and wallpaper patterns, hexagons on my bathroom floors in the sanatorium.

Since hexagons can tile a flat surface, it must be possible to envisage a two-dimensional universe shaped like a hexagon that repeats infinitely in all directions (Figure 12.6). A visual image of the topology of this space is a little more accessible if we artificially bend the hexagonal tile into three dimensions and glue the sides together. What we'd find is another way to make a torus.

What condenses from this trial and error is a brief manual to topology building: cut a fundamental shape from space and identify the sides to make a multiconnected and edgeless space. Equivalently, we can tile space with these shapes and see the spectrum of ghosts emerge.

We followed this prescription to make five flat spaces, the infinite plane, the cylinder, the torus, the Möbius strip and the Klein bottle, all of which are flat. We could apply the instructions to an octagon. Take a flat surface, cut out an octagon and identify the sides (Figure 12.7). However, if we tried to tile a flat surface with octagons, the tiles would overlap and could not smoothly cover the surface. If we glued the

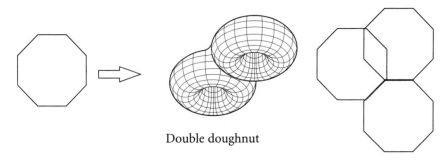

Double doughnut

Figure 12.7 *A double doughnut can be made from an octagon. Although the octagon cannot tile a flat surface, it can tile a negatively curved surface.*

octagonal tiles together without forcing them to try to lie flat, they'd curl up and try to make a negatively curved shape, which is a hint towards the truth. If instead we cut out an octagon from a negatively curved space, obeying Riemann's geometry as we did so, then the interior angles of the polygon would narrow on the negatively curved surface. This narrowing allows us to tessellate the negatively curved surface with octagonal tiles without gaps or overlaps (Figure 12.8). We can cheat and bend the octagon into three dimensions, glue opposite edges and verify that we have made the double doughnut. A two-dimensional inhabitant cannot visualize the space from a third dimension, but from the

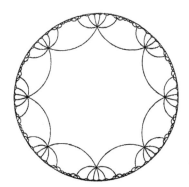

Figure 12.8 *The intuition for tiling negatively curved space with polygons comes from remembering that the straightest lines on the curved surface look to us like arcs. As a result, the triangle drawn on the hyperbolic plane narrows. If the triangles are drawn just the right size on the negatively curved surface, they can just manage to squeeze together with the right number and orientation of tiles to fill the plane without gaps or overlaps.*

restricted two dimensions he could see a spectrum of images fill the sky with patterns that reflect the octagonal shape of the peculiar little cosmos.

Tessellations of negatively curved surfaces were used by the Dutch artist M. C. Escher in his quizzical drawings. But then lots of great artists played with geometry, like Picasso, Juan Gris, Malevich, Salvador Dali, Mondrian, even the eccentric Marcel Duchamp. I wonder what they would have done with all of this information. If I had them here I could condense these ideas into a brief artists' manual to topology, a three-step set of instructions to understanding finite spaces: (1) Choose the geometry: either flat, positively curved or negatively curved. (2) Cut a fundamental shape from this cloth. (3) Tile the geometry with identical copies to simulate the finite space.

And now the conference is starting. I wear my mittens during the lectures earning me some taunts, but I don't mind.

# 13

## WONDERLAND IN 3D

**23 FEBRUARY 2000**

Up a dimension to three. I moved around in all three dimensions today, as we always do, but only a little more so. I was taken north through Moscow to the airport (or was it south?), then up in a plane, then west through the air to the West and then down again on to London. Here we are in three dimensions. Unable to bend a space into four and look at it. Unable to naively see the handles and holes. But we can use two important techniques that we used for the two-dimensional compact spaces. We can identify the faces of three-dimensional shapes or we can use rules to tile space.

If the universe is in fact finite, then even if we travelled in as straight a line as space would allow, never turning or stopping, we might end up where we started. We could leave the earth and travel to what appeared to be a distant planet, only to discover that the planet was our home the earth. A good twist on the premise of *Planet of the Apes.*

It is extremely difficult to visualize how three-dimensional space could be smooth, edgeless and compact. We could take a three-dimensional cube with a flat geometry, for instance, and imagine gluing the faces together to make a space compact. But we'd be thwarted since we'd need to bend the cube into four dimensions when we have access to only three (Figure 13.1). The tiling picture becomes very useful when we promote ourselves to three dimensions. In the tiling picture the compact space is represented as an infinite collection of identical copies of the cosmos that are fitted together face to face (Figure 13.2). The faithful tiling picture allows us to visualize life in a compact three-dimensional space without artificially bending the space and falsely distorting the geometry.

A three-dimensional compact space can be made by gluing the faces of polyhedra. However it is not possible to draw compact three-dimensional topologies because it would require four dimensions for the embedding.

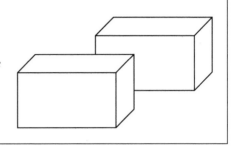

Figure 13.1 *Three-dimensional compact manifolds.*

If we lived in a finite cube, the universe would still be edgeless and smooth. We'd notice some strange things about the space in which we lived. If we travelled far enough east we'd exit the face of the cube only to re-enter from the west. We would not perceive an edge – there is no real edge, just a continual wrapping of the smooth connected space. If we travelled far enough away from the earth towards the north, we'd find ourselves moving towards the earth from the south, and far enough up, we'd find ourselves moving towards the earth from below. We can imagine tiling three dimensions with identical cubes. Each cube represents the universe. If the earth is in the centre of the cube, we would see the earth in the centre of every copy, filling three dimensions. If we travelled away from the earth towards the north, we'd eventually approach the earth again from the south, and if we passed the earth this time and continued north, we would once again approach the earth from the south. If we could travel forever in the northward direction, we would continue to fly past the earth each time we crossed the finite breadth of space.

This space has handles which we could not see any more than the

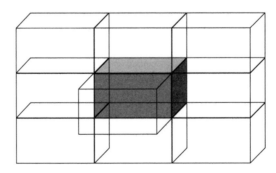

Figure 13.2 A *tiling of flat three-dimensional space with rectangular volumes.*

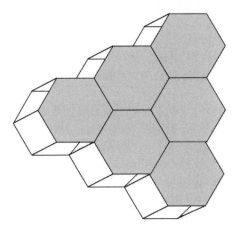

Figure 13.3 *Another compact three-dimensional space can be built out of a hexagonal prism and visualized as a tiling of a flat three-dimensional volume with hexagonal prism tiles.*

inhabitant of the torus could see their handle. We could detect the handles with irreducible loops. Hypothetically, a ball of string could be tied to the earth and unravelled behind a rocket ship that travels north in a straight line. When the rocket reaches the earth, approaching from the south, the two ends of string could be tied together. The string forms a closed loop around the compact direction. An experimentalist could pull the string, hand over hand, to see where it ended. The knot would recede from the earth until the experimentalist discovered precisely the same knot approach in front of him. There would be no way to unwrap the string without cutting it. Two other loops of string could be wrapped in the other two directions. These irreducible loops help to identify and characterize the shape of space.

Since tiling is equivalent to gluing the faces of a shape to render space compact, we can visualize different topological spaces this way (Figure 13.3). Just as we can tile a flat two-dimensional floor with squares and hexagons, we can tile a flat three-dimensional volume with cubes and hexagonal prisms. Just as we can tile a negatively curved two-dimensional surface with octagons, we can tile a negatively curved three-dimensional volume with polyhedra such as an icosahedron, a polyhedron with twenty triangular faces (Figure 13.4). Icosahedra could not fit together evenly to tile a flat volume but using our knowledge of non-Euclidean geometry, we can tile a curved space with the icosahedron. On a space with negative curvature, we could draw icosahedra exactly the right size, so that the corners narrowed and the faces bowed until they fit together perfectly and we could cover the curved volume evenly with these Platonic tiles. We could also tile a positively curved space with icosahedra, with different rules for laying down the tiles edge to edge.

The way these tiles fit together can get complicated. If we lived in the compact icosahedron, then as we left the earth travelling in a random direction, we would find ourselves approaching the earth from another unexpected direction, and if we continued, we would pass the earth again, only to find it coming closer in yet another direction. As we moved forever around the finite space, the earth would loom ahead of us, behind us, to the left or right of us, but there's only so far we could ever get from home.

We can make a catalogue of compact spaces with our manual: (1) choose a geometry (flat, positively curved or negatively curved), (2) cut out a fundamental shape and (3) find ways to identify the faces. I could bombard you with three-dimensional athletics, with lists of spaces to rotate, glue, flip, bend, twist and identify. My neighbour Ben coined it 'intergalactic origami'. Occasionally I see someone in the back of one of my lectures trying to fold paper into compact spaces then waving the origami at me for approval.

The catalogue of idealized three-dimensional finite spaces is long and the list is unfinished. The mathematicians have not fully classified the spaces, but they've made remarkable discoveries trying. We can generalize the sphere, the torus, the Möbius strip, and all the other two-dimensional

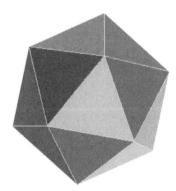

Figure 13.4 *The Best space made from a regular icosahedron. The figure is taken from the SnapPea program written by Jeff Weeks for computing three-dimensional hyperbolic topologies.*

spaces in this way. In addition to their three-dimensional generalizations, there are an infinite number of unforseen possibilities.

Flat topologies are the easiest to manipulate. The cube[1] described above is a three-dimensional generalization of the flat torus, called the hypertorus. The rest of the flat topologies can be made by imposing different tiling rules on the cube or the hexagonal prism. We could tile the cubes so that every other cube was upside down. Consequently, as we flew north, we'd find the earth inverted on every other pass through space. We can twist and glue and catalogue all of the possible topologies that emerge. In all there are seventeen flat spaces. The list of positively curved topologies is infinite, but it is countably infinite. By this I mean that we have a simple counting prescription for generating all of the possible positively curved spaces. Some infinities are larger than others. There are more negatively curved spaces than positively curved spaces. There has been great progress this century in classifying the compact, negatively curved spaces, but they continue to resist complete classification. This means that even the mathematicians haven't finished with them yet.

This has been an unsatisfyingly incomplete course through topology. There is so much more. For instance, this kind of construction based on gluing polyhedra is just one way the topologists discover new connected spaces. There are other methods such as Dehn's surgery which is too complicated to do justice to here, but simplistically the method starts with a space with cusps which are horn-like corners. Toroidal shapes are drilled out of the cuspy corners and replaced with new solids. In this way topologists are still discovering new shapes for space, each with unique properties and potentially unique cosmological signatures. If we do live in a finite space, we can only live in one out of the infinite set of possibilities. Which one? Now that's the question.

Although this is not a systematic or complete review of the mathematics, I do think this is enough information for our purpose, which is to understand the spectacular images that would fill the sky of a compact universe. Still, a few last pieces of information will help in the application to cosmology.

The great contemporary mathematician William Thurston, who won a Fields medal for his bold insights, realized something tremendous. It is clear that while curvature and topology describe different facets of a space, there is a link between the two. Topology is restricted by curvature, though not completely determined by curvature. We know that a

---

[1] More generally, the cube can be replaced with a parallelepiped.

Figure 13.5 *The Weeks space is built by identifying the faces of this fundamental shape. The figure is taken from the SnapPea program written by Jeff Weeks for computing three-dimensional hyperbolic topologies.*

given tile must be drawn precisely the right size on the curved surface in order for the interior angles to adjust and allow a complete and even tiling of the surface. One would not normally think of area as a topological feature since I can change the area smoothly by stretching or shrinking, but the area of the surface becomes inextricably linked with the topology. Something even more confining happens in three dimensions. Not only the area, but all lengths, such as the length of the shortest path around the compact space, become fixed by topology. The proof of this three-dimensional stiffness is known as the rigidity theorem.

The theorem is important to cosmology. It implies that at least some spaces are naturally small. If we happen to live in one of these, then we have a hope of seeing the effect of the shape of space on the cosmos. If all lengths and volumes are rigidly fixed in three dimensions, then all compact topologies can be classified entirely by ordering them in terms of ascending volume. The smallest negatively curved space known is the now famous Weeks space, named after my friend and collaborator Jeff Weeks (Figure 13.5). Weeks was Thurston's PhD student and displaced Thurston's famed space with his own for the coveted title of smallest negatively curved manifold known. There may exist even smaller spaces, but they haven't been found yet.

The smallest spaces are the most important to cosmology. If the volume of space is huge, then we won't be able to see far enough into the universe to perceive the topology. We would not be able to tell the difference between a finite and an infinite space. If instead the volume of space is small, then we just may be able to see far enough into the universe to discover the shape of space. Consequently these idealized small spaces, such as the Weeks space and the Thurston space, are used to theoretically model different cosmologies. We then compare these theoretical models with astronomical observations to see if we can locate ourselves in the cosmos.

If space is finite, then we will be able to place ourselves in relation to the handles and the cuspy corners, if there are any. Topology amends the Copernican principle. While we might not be central to the cosmos, someone is. At least there is a definable centre in a finite space, unlike an infinite space, and in principle it could be located on a cosmological map.

### 25 FEBRUARY 2000

Dad has set me straight. He has explained the subtle yet crucial difference between sanatorium and mental institution. I was in a sanatorium in Moscow, a place to rest, like a spa, not an asylum. I wonder if they make such a distinction in Russia. Russia on the earth, the earth in a giant polyhedron, a three-dimensional Platonic solid.

Do I really think we live in a perfect three-dimensional compact polyhedron? Not exactly. It's more that I think space is finite and the perfect polygons are mathematical idealizations that allow us to pursue the implications either to the point of absurdity or to the point of discovery. Using the ideal forms, we can devise thought experiments to determine how we would know if the universe was finite.

A good thought experiment is to imagine inflating a giant balloon throughout space, an image I borrow from Jeff Weeks (Figure 13.6). Initially our elected observer will see a confounding pattern of ghost images in what appears to be infinite space. She can inflate a balloon in her small compact geometry and she will see all of her ghost images inflate balloons. The balloon can inflate until it exceeds the size of the space and the elastic membrane begins to press against itself as one point in the balloon fills out far enough to collide with another. As the balloon presses and presses, the faces of the fundamental shape become clear and she can thereby determine the shape of her universe.

If the universe were finite but big, say bigger than a few hundred light years across, then our elected observer would not be able to blow up such a colossal balloon, nor would she live long enough for the light of her reflection to travel around the space (Figure 13.7). She would see no copies of herself, but she could see many copies of her solar system, which has been around for millions of years. She could watch the solar system form from dust and coagulate into the star at the centre with its collection of planets. Or the universe could be bigger, big enough to fit a

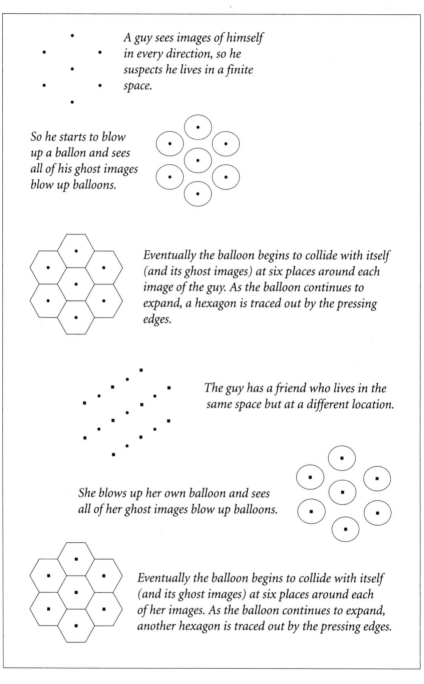

*A guy sees images of himself in every direction, so he suspects he lives in a finite space.*

*So he starts to blow up a ballon and sees all of his ghost images blow up balloons.*

*Eventually the balloon begins to collide with itself (and its ghost images) at six places around each image of the guy. As the balloon continues to expand, a hexagon is traced out by the pressing edges.*

*The guy has a friend who lives in the same space but at a different location.*

*She blows up her own balloon and sees all of her ghost images blow up balloons.*

*Eventually the balloon begins to collide with itself (and its ghost images) at six places around of her images. As the balloon continues to expand, another hexagon is traced out by the pressing edges.*

Figure 13.6 *The edges of the tile are meaningless. No observer perceives an edge as they move through space. They could find ways still to deduce the shape of space.*

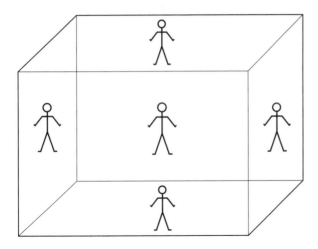

*A solitary observer in a small space sees copies of herself in all directions.*

*The universe looks like a hall of mirrors.*

Figure 13.7 *View of a compact flat universe.*

galaxy, a collection of billions of suns and planets. She'd have to look for a pattern of galaxies in the sky that could be ghost images of her own galaxy (Figure 13.8). The bigger the space, the older and bigger the relics she would have to look for to determine the geometry of space. If the space were so huge that not even galaxies are ancient enough for their ghost images to wrap around the cosmos, she'd have to look back to the beginning, to the hot and cold spots in a bath of radiation left over from the big bang. This is the game we've been playing in cosmic topology.

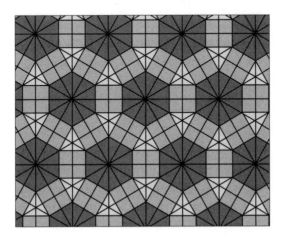

Figure 13.8 *A compact cosmos creates a kaleidoscopic pattern of ghost astronomical images.*

Can we observe the extent of the universe by looking for a pattern in the universe's spots?


**27 FEBRUARY 2000**

I've lived in at least ten different cities in four different countries. I've been through Jerusalem, Europe, Japan, some of the UK, some of the US. I'm *tired*. So tired.

Moscow was the turning point. I came back to find Warren catatonic on the couch. He hadn't spoken to another human in a week. His eyes were wild but unfocused and he didn't even get up to greet me. He looked like a man who had just climbed out of the woods. His beard had overgrown his mouth in a wiry trap. I tried to console him, lure him back to this planet, but it was the beginning of our end. I could see our end coming, though it took six more months to blossom. Incubating for the winter.

We don't survive. By May I will move to London by myself and it starts over. My life starts over. I didn't know this yet, but I have the acuity of hindsight now. I can sneak back into the early dates and edit my thoughts. Would my friends have made the same choices, I wonder? Will their relationships survive?

I'm telling parts of a story I don't really want to reveal. This is a story I'd rather not tell, but the terror of having life come and go with no record impels me forward. If I don't, my stories will die before me, victims of an incompetent memory. Not that I'm feeling dramatic.

# 14

## MIRRORS IN THE SKY

**3 MAY 2000**

I remember watching you on the couch silently reading for hours, absent-mindedly wiggling your toes. I didn't know how you could do it, read so much. The compulsion hit me suddenly like a swiftly descending disease. I inherited your penchant for reading as surely as I inherited your eyes and hair. I haven't been able to stop. If I'm stationary for more than a few seconds, my face is in an open book. I'm not cramming my brain with knowledge by reading so much. I'm humming along to a tune. Listening, absorbing, not studying. Massaging my inner workings. I've a list of over twenty books I've read in the past four months. I actually think it's more in the thirties, but I can only remember about twenty-five titles. Some of them were admittedly not very good. I have no quality control when it comes to obsessive bouts of reading. I will happily read any old schlock, but there was some brilliance in my list of novels too.

A warm quiet envelops me when I read. I've been carrying silence around me like a big cylindrical wrap to heal my wounds. It's like carrying a shell to my ear to hear the soothing ocean of my own blood. I admire the authors' devotion to make something beautiful and useless. I can defend my most abstract research the same way, but scientists don't appreciate useless. Look how beautiful and useless, I feel like saying. Define useless anyway.

Is this math useless? Should I stop staring into space, which manages to be disarming by not staring back – a delicious emptiness with a few bright sparks too distant and too old to return a threatening glare. Shouldn't I be more concerned with my ordinary life? Or maybe what I really should say is: shouldn't I accept the ordinariness of life? My friend

Prudence keeps asking me, does it affect you, what you think about all day? Does it? I'm not sure. Does it change your worldview? I don't know. How do you carry it with you? How can I know?

I can answer the first question, but none of the others. The math at least is not useless. We just discover things in a peculiar order. We didn't always know that topology had anything to do with the universe. When I first started asking the question, I posed it in all the wrong ways. I wondered about compact hyperbolic manifolds, chaotic flows, homotopy groups and homology. Who is to say the perfection of the circle is more complicated than the infinite piece of string? Mother nature doesn't care if the math is hard for us. She cares if it's easier to make a finite universe or an infinite one and then takes the path of least resistance. Nature always does what's easiest. Water rolls down hills not up them. Materials disintegrate, atoms don't spontaneously collude to form a chair. Entropy increases. Hang the math.

So what did she choose for the global shape of space? Which is it, and why that shape? I can argue on aesthetic grounds that the universe should be finite, but we'd all be a lot happier if we had a prediction from a fundamental theory. We simply don't have such a prediction because we don't have such a theory. Remarkably though, we might be able to look out into the vastness of space and observe if the universe is small enough to reveal her size.

If nature had made space as small as a room, we'd always have known the world was finite. If I stand in the middle of a compact universe with a light bulb that never burns out and space is as small as a room, the light from the bulb will transit around the space. Some of that light will make it back to me after winding once around the space. Some of that light won't get back to me until winding a few times around the space. Some of that light will take nearly forever before it just happens to scratch past me again. If I wait long enough, I'll see a copy of myself in all directions, so that the space is filled and bright with ghost images of me and my light bulb at different ages, a haunting kaleidescope of images (Figure 14.1). I could watch myself growing up and growing old. There'd be no need for photo albums, I could watch your life before me beyond the images of my own.

The finite universe exacerbates a paradox due to a German astronomer, Wilhelm Olbers, in the 1800s. Olbers noted that if the universe is dense with stars, it is surprising that the sky is as dark as it is. If we were to blot out the sun and turn off all of our electricity and all sources of light, the sky would be fairly sparsely dotted with individual

Figure 14.1 *A hall of mirrors view of a compact space. The further away the ghost image, the longer it's taken light to reach the observer.*

stars. The sparseness of the stars is mysterious if the universe has lived forever and is infinitely filled with stars, since then we should hit a star in every direction we look. Like looking into a thicket, we'd see a blanket of shrubs, or in this case stars. The sky would be a bright wall of light where we wouldn't see the trees for the forest. The resolution lies in the finite age of the cosmos and the relative youth of its stars. Even if the universe stretched infinitely far, stars don't burn forever. They ignite and burn out, so that at any one location in the cosmos we can only receive light that has managed to reach us from its original source.

The finite universe worsens the situation by trapping light in the compact space forever. Light continually traverses a finite cosmos inevitably returning near any given point in the universe. We could be blinded by the light of one star, one eternal light bulb for that matter. But nothing lasts forever. Stars die, light bulbs expire. Not even a finite universe will be blinded by the light of its stars. Maybe someone could estimate the size of the universe from these simple principles of burning stars and their abundances, but so far no one has tried.

I know, of course, that the universe is bigger than my workshop, since I can travel across London without ending up where I started. I know the universe is bigger than London, since the train I take to Manchester gets me there and doesn't deposit me back in London, barring crashes, cancellations and other recent fiascos. The universe is bigger than the earth, since I can get into a plane, fly in a reasonably straight line and get out in Paris. But if I flew in a rocket out into space in a straight line, never turning or stopping, would I go on forever or would I follow the wrapping of space to see the galaxy I left receding behind me approach in front of me?

I know it's bigger than the solar system because we don't see copies of the solar system. I know it's bigger than the galaxy because we see other galaxies out there, so we know it is at least big enough to house millions of galaxies with their billions of stars. But maybe some of these other galaxies are really our own galaxy. Maybe one of the ancient galaxies we see in the distance is really our own galaxy at an earlier age. The light having taken millions of years to get here, we could see our past unfold. Or maybe it is even bigger, so that light hasn't had a chance to make it around the space, not even in 14 billion years. As with an absurdly long movie that outlives its audience, we would not live long enough to see any action.

## I JUNE 2000

Pedro and I are as thick as thieves. He and Joe Silk have organized a conference at Oxford and I'm to speak at a joint session of a cosmic background radiation meeting and a particle physics meeting. Pedro assigned me the title 'How Big is the Universe?', which is just as good as the ones I usually use, which vary from 'Is the Universe Infinite?' to 'Is the Universe Infinite or Just Really Big? We sit together in the front row as Pedro chairs the sessions and we exchange silent conversations composed of gestures and expressions that we've learned to read too well. Only rarely do we have to resort to the written note.

I don't really propose to know how big the universe is. I only tell my jaded audience ways that we've devised to look. I tell them academic anecdotes and try to dazzle them with tricks of the trade.

As a model, we can pretend for the sake of argument that we live in one of the ideal Platonic forms. The topological connectedness acts as a lens, sending light on continuous paths through the connected space (Figure 14.2). If light exits one face, the connections in space will determine from which direction the light re-enters. A pattern would emerge in the distribution of galaxies if many of these galaxies were really ghost images of others. The pattern would reflect the pattern of gluings of the fundamental shape.

One consequence could be a pattern of images of galaxies, some of which are ghost images of the others. People have managed to use statistics of observed galaxies to search for spatially significant patterns. So far this is a brutal task and the practitioners have the woes of low numbers. Galaxies have not been around forever and there are only so many that we can see at such great distances. Also, being unable to anchor the real

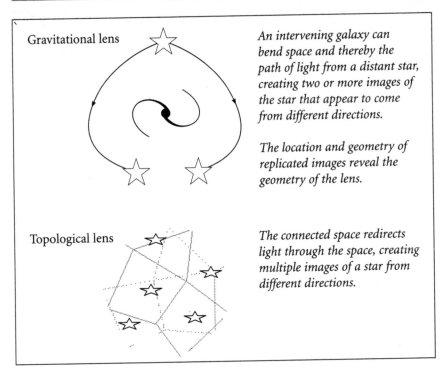

Gravitational lens

An intervening galaxy can bend space and thereby the path of light from a distant star, creating two or more images of the star that appear to come from different directions.

The location and geometry of replicated images reveal the geometry of the lens.

Topological lens

The connected space redirects light through the space, creating multiple images of a star from different directions.

Figure 14.2 *Topological lensing is akin to gravitational lensing.*

identity of a galaxy as it changes and evolves is an impediment. Trying to identify a galaxy at different ages is as difficult as trying to match baby pictures up with their owners in a room of the elderly. Some researchers have tried to tackle this difficult problem by defying the obstacles with sheer cleverness.

Maybe galaxies happen to lay themselves into specific patterns for other reasons. We do see vast voids and huge structures of connected islands of galaxies and clusters. We still don't know why large galactic structures pile up this way. If we did a survey of galaxies along a narrow line of sight, we might intercept galaxies that cluster on huge bubbles. The effect would look like a periodic repetition of galaxies, which we might mistake for topological clone images. So observations of galactic patterns may lead to ambiguous conclusions.

There is another way to look for topological lensing that does not depend on evolutionary effects but instead relies on one of the oldest artefacts of the big bang, namely the cosmic background radiation. This cool bath of radiation left over from our genesis is the oldest witness to the universe. Minute hot and cold spots provide a fossil record of our earliest history. In principle, we can observe a pattern in the universe's spots that traces out our geometry. These observations are just within reach as we await the launch of capable satellites within the next couple of years. Without leaving the earth, through a kind of cosmic archaeology, we could see if the universe is finite.


## 5 JUNE 2000

I'm on the coffee-grey washed streets of Hackney, in the East End of London. All the colour scrubbed away with the rain that threatens to leave but never does. Car lights blink the way, guiding me to the curb. This is the first time I shop without him. I step on to the curb and past the bikes. I step up to the glass and past my solitary reflection. I drift through the limp vegetables and brown cans, placing cautiously little in my dirty orange plastic basket. There must be a way home from here. This is the turn in the maze I chose and there must be a way home from here. Home. Home. Home. Home is where the hat is. Wherever you go, there you are. Here I am. London.

London almost killed me those first two months. I phoned Leslie or Stacey every day sometimes just to have them hear me hyperventilate, and when they'd coaxed my breathing back to normal I could disengage. I guess that's what sisters are for. That's what they reassured me anyway. It's hard to believe that I gave public lectures in my state of complete monotone and mental retreat. Just stand me up in front of an audience with a few transparencies in my hands and wind me up – a scientist's song and dance and leaps through hoops of flame, entertaining myself more than anyone else.

Lifting my head from the fog, I am surprised to find myself in a new home in a new city in this foreign country. I was relentless in my search for a warehouse space and I've found one. The first person I met in London, Mark Lythgoe, by coincidence or serendipity happens to live across from the industrial premises that I now call home. He stares at the rusted edifice from his side of the canal and wonders if I have gone mad. Maybe I have. Before the move I called Alene every night at around 4 a.m. when I couldn't sleep, 8 p.m. Los Angeles time, and she'd talk calmly on the phone, make lists of pros and cons to the warehouse move. We'd decide rationally that I should make the move, live large. Then I'd call her again the next night and she'd calmly go through the whole process again as though she hadn't just done so last night. We did this for about a week and then I moved in.

It's not fair to call my place an apartment. It had no bathroom, no kitchen, no walls. It was an empty shell. Chris Isham has named it the boiler room, sight unseen, and it's a very accurate description. I love it. Sometimes the drilling is so loud from the manufacturers above me that the boiler room rattles and dust and paint chips sprinkle the room. I pretend it's celebratory confetti. I bought a bathroom over the phone, the Milan suite in white, only £149, had it delivered and Len, the trusty on-site builder, installed the loo and built a partition. The partition doesn't reach to the ceiling and can't be 6 feet high. It looks a bit like a large doll's house. I do not own a home, but I own a bathroom. A bed and a bathroom, which is all that I have to fill the 838 square feet, not enough to dampen the echo.

Pedro is mortally offended that I don't have a television and chastises me daily. He can boast about his avarice for television with the confidence of an Oxford faculty member in Astrophysics. He's taken to turning on the television when he calls on the phone and narrates for me a scene from *NYPD Blue* or *Big Brother*. Nothing's happening in the NYPD right now, but he'll fill me in when there's a lot of flurry.

Pedro works on the more reliable aspects of the geometry of the universe. He's bubbled up through a vast industry of talented people who try to decode from the cosmic background radiation the curvature of the region of space we can see. We're just observers in this universe and we cannot see infinitely far out into space. We can only see as far as light has travelled since the beginning of the universe and really not even that far. The energy of the universe's birth was so immense that we can only speculate about the first instants when quantum gravity dominated and spacetime itself was ill-defined. But we have a great deal of confidence about the fairly late creation of the cosmic background radiation. The light flooding all of space was released from its clanging prison of charged particles at least 10–15 billion years ago. This light from the big bang is unleashed into space to move fairly unhindered for the rest of time. The bath of radiation floods the cosmos, a reminder that the big bang was not an explosion rushing away from a centre but rather that the centre is everywhere. The centre itself is becoming more diffuse as the universe continues to expand and cool. Stars form, explode, form anew. Planets coagulate from the soil of generations of stars' debris. People take their time to evolve, build telescopes and satellites and point them in the sky to see where they came from. There it is, the cosmic background radiation, awash with minute secrets about time before organic life.

The cosmic background radiation tells us about the largest attributes of the universe and indicates that the universe is smooth and evenly curved, leaning towards flat. There's something delicately unnatural about how smooth and evenly curved the universe appears to be. If the universe is globally curved then it is only slightly so, with the curvature just perceptible on the limits of the horizon.

Cosmologists are often concerned with naturalness. If something seems heinously unnatural, we'd seem fatuous attributing it to nature. If we came across a sandpile in the desert in the perfect rendition of Elvis, we'd be unlikely to suggest it as a sculpting of random winds. Similarly, if the universe is cast in a suspiciously perfect rendition of some Platonic form, we won't be inclined to attribute this to the unlikely conspiracy of random forces. Cosmologists want a causal explanation.

As it happens, in our current cosmological models, it is very difficult to make a universe that looks like ours. The standard model of cosmology derived from Einstein's theory tells us that there are essentially three choices for our local environment. The universe appears to be either everywhere flat, or everywhere positively curved, or everywhere negatively curved. I don't really mean *everywhere*. I mean only as far back as we can see, which is at least back to the time when the primordial radiation stopped scattering off the hot primordial soup. Beyond that, the universe could be lumpy and bumpy with impunity and we'd be none the wiser.

What needles us is that if the universe is slightly positively curved, it is inclined to recollapse long before galaxies, planets, telescopes are formed. In other words, long before we'd be here to observe it. But here we are. If the universe is slightly negatively curved, it is inclined to expand so quickly that we'd live in a colder, more sterile cosmos than we observe. If the universe is truly flat, at least it will remain so. But if it was even slightly off from flat, it would collapse or cool in short order. So it is unnatural that we should live in a cosmos where we just happen to be able to barely see curvature when this is a terribly unlikely turn of events. Inflation is motivated by concerns over naturalness and most sound inflationary scenarios explain why the universe should appear flat. In short, if inflation occurred, the universe blew up incredibly rapidly in the very early moments, so that any curves and lumps are pushed far out of view, leaving the relatively small patch we occupy essentially flat and smooth. For this reason many theorists argue it is more natural for the observable universe to be flat than for it to be nearly flat.

A similar naturalness argument mars topology. If the universe is finite, why should it just happen to be within view of the observable horizon? Why isn't it much much smaller or much much larger? I don't know.

The near flatness of the observable universe is a subtle and delicate problem that I won't do justice to here, but I'll share the observation that after all we can almost see the curve of the earth. Why should the curvature of the earth be just on the verge of detectability, so that it appears flat from the ground but just beyond the horizon turns out to be curved and compact?

There are Darwinian reasons for why humans are the size they are relative to the curve of the earth. If we were too big, we'd be unable to handle our skeletal structures and if too small presumably something else bad would afflict us. In theory, we have evolved to a fairly optimal size given our other characteristics, such as a meaty cerebral cortex and some physical dexterity.

Could there be Darwinian explanations for our size in the cosmos? Some reason why we evolved to be able to just barely see the curve of space? There could, and these explanations range from anthropic principles to ideas of Lee Smolin's on natural selection. I think Lee would argue against the former and on behalf of the latter.

The anthropic principle argues that we live in a universe with these conditions because they are the only conditions that could support life. The mass of the fundamental particles, the strengths of the fundamental forces, the shape of space, are all just right for a universe hospitable to life. The weakest form of the principle argues that the conditions could very well have been different, and may be very different in distant regions, but we simply would not be here to ask the questions. The strongest form of the principle asserts that the universe was tuned to spawn life – an assertion rife with allusions to a power beyond nature.

Some models of inflation tie in with the anthropic principle. Eternal inflation promoted by André Linde leads to a ginger root of inflationary universes bubbling off forever into the past and forever into the future. In this model there was no beginning to the universe. A big bang is just the period of heating following inflation when all of the energy trapped in the cosmological constant is released as heat and a smouldering bath is generated. For all we care, this is the primordial soup that cools and encases us today as the cosmic background radiation. We can't see past this blinding flash of light and are causally quite separate from the other patches that have also endured an end to inflation, a burning release of heat and a familiar evolution. Familiar but maybe not identical. It has been suggested that the values of the fundamental constants are different in different bubbles. The strength of gravity is different, the mass of the proton is different, the values of the charges, the things that organize the world as it is. In another patch, with different fundamental constants, the world would be organized utterly differently. If things weren't so tuned, there would be no primordial nuclear fusion to synthesize the common elements, no formation of galactic structures, no organic matter. There'd be no life, no us to even ask the questions. This fairly light form of the anthropic principle says that we happen to be here to ask cosmological questions because the values of the fundamental constants are what they are. This isn't the most comforting explanation. We frankly aren't good enough at physical cosmology to truly predict what kind of universe would be generated by different values of the fundamental constants. There could be unforeseen structures, unforeseen life. But fair enough, not us.

Lee Smolin suggests a cosmic natural selection in his book *The Life of the Cosmos*. He hypothesizes that in the centre of each black hole could be a new universe separated from our own by the black hole's horizon. In each universe the cosmic conditions may be slightly different from our own; different geometries, different particle masses, different interaction strengths. Slight differences in these elementary properties will change the world as we know it. Most importantly, he argues, the production of stars and their demise will be altered. The elementary properties are like cosmic genetic information that will get passed on to the next generation of black holes and the subsequent universes ballooning from their centres. Nature will statistically select the conditions most favourable to the production of black holes, since the more black holes there are, the more universes are born and so on until it becomes extremely likely that we live in a universe with precisely the optimal cosmic genetic information to produce black holes. If we were better at physical cosmology, we could test Smolin's hypothesis by determining if in fact the universe we live in has optimal conditions for the production of black holes. Too many factors are at play for us to be able to predict which conditions optimize black hole production, and Lee's idea is likely to remain untested for a long time.

I'm not a big fan of anthropic arguments, but we live in strange times. We have to keep our eyes open and our minds limber. Truth is, we just don't know the curvature of spacetime. And we don't know how big it is. As long as satellites are being launched to finally determine the largest curves in space we can see, we might as well piggy-back and sneak a deduction of the topology too.

If I knew nothing about topology, I would *assume* that the universe was infinite if flat, finite if positively curved and infinite if negatively curved. Each of these corresponds to the assumption of simple connectedness. None has handles or holes. But we can impose a topology without altering the curvature and can more or less preserve the standard cosmology of both our early history and our ultimate destiny. If the universe is compact and has handles, these features will never change during the adult lifetime of the cosmos. Although relativity makes predictions about the curves in space, it cannot anchor the topology. But we can, despite our ignorance, try to look and see. We owe this to ourselves as a defence against fanaticism and doctrine.

# 15

## HOW THE UNIVERSE GOT ITS SPOTS

**11 JUNE 2000**

I like to write from memory, as imperfect a record as that provides. My memory has been kind to Warren. Sometimes I wake up and think I should edit him from these letters. The last time I saw him, he sat there frantically tracing triangles between his fingers. It seemed a hectic illustration of our lives together, weaving us in and out of our past and future to the tangled vision of who we are now.

I was at a Christmas party when someone asked, do you know what they call a musician without a girlfriend? Homeless, I guessed correctly. I can't help but chortle, but with no small dose of tenderness. I did hear he was living in a tent, but he probably loves it. Man of simple means and few possessions. I was back in Brighton for an afternoon and admit that as I climbed over the lean limbs of some homeless guys on the street I looked at them long and hard. None of them was him. Maybe he's flourishing without me. Finally able to get on with his own life instead of struggling in the wake of mine.

Musicians and mathematicians are supposed to have a lot in common. From the outside I suppose it must have seemed we had nothing in common except for a sense of humour and so much other good stuff you can't easily identify.

Music and math are often complementary. A mathematician, Marc Kac, asked if we could hear the shape of a drum and my colleagues, Neil Cornish, David Spergel and Glenn Starkman, knew that we were asking a similar question when we asked if we could see the shape of the universe. If we pluck a string it vibrates at a certain frequency, creating waves in the air. When the waves strike the mechanisms in our ears, we hear sound. Higher-frequency waves with shorter wavelengths sound higher in pitch than lower-frequency waves with longer wavelengths. Our ears and brains collaborate in response to certain superpositions of frequencies so that, while we may be lulled by the music of a cohesive song, we can still separate a cello from a violin from a tin whistle. Electronic equipment in studios outperforms our ears and can technically separate the different sounds on the basis of their frequencies, separating bass frequencies from treble.

The hot and cold spots in the sky encode a kind of frozen moment in a musical score. If space were perfectly smooth and the cosmic background radiation reflected this perfectly, the light filling the cosmos would primarily ring at one wavelength, one note.[1] Space does appear to have been very smooth when the primordial radiation was first freed and the cosmos was about 300,000 years old. However, we know from the Heisenberg uncertainty principle that nothing can ever be perfectly smooth or perfectly still. Consequently, small perturbations on the otherwise smooth space are planted through quantum processes in the early universe. These perturbations create mild hills and valleys on space. These are like the waves on a drum. The hot and cold spots are etched into the cosmic background radiation as light climbs out of these hills and valleys. At the time of last scattering, this permeating score is frozen into the cosmic background radiation, only lowering in key as the cosmos expands. The very first ambient music. If the universe is finite, this score is not just cacophony but has natural harmonics and imprinted patterns – melody. This is what we search for.

If the fluctuations were entirely random in the early universe, we would expect to see a spectrum of hot and cold spots approaching white noise: that is, a random and structureless superposition of temperature spots of any size. If instead the fluctuations are confined to a finite box, namely a finite universe, the shape and geometry of that box will mould the spectrum of fluctuations. By looking for the discrete harmonics of the different shapes of space, we could determine if the universe was compact and connected, we could hear the shape of the drum. If we cannot see the discrete harmonics, we'd have to conclude that the universe is too big to see all the way through it, but we may never be able to assert that the cosmos is actually infinite.

---

[1] The spectrum of radiation is not really monochromatic. More accurately the radiation spans the thermal spectrum originally derived by Bohr. The thermal spectrum is highly peaked at a given temperature which means that most of the primordial radiation has an energy and frequency proportional to this temperature. The temperature of the radiation thereby fixes the loudest note.

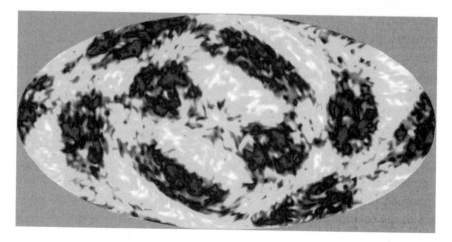

Figure 15.1 *An elliptical projection of a computer-simulated sky. The pattern shows the hot and cold spots in the cosmic microwave background in a hypothetical universe with the shape of a twisted hexagonal prism.*

This means that if we lived in a small hexagonal prism, we could actually see the shape of the prism in our snapshot of the sky. We calculate the natural harmonics of a given shape and then simulate our calculations on computers to generate images like Figure 15.1. This is a theoretical prediction for a map of hot and cold spots in a universe the shape of a hexagonal prism. The map is an elliptical projection of the spherical sky, analogous to an elliptical projection of the globe. The presence of hexagons is undeniable in the figure, as are the twists in the prism direction. Is this our sky? Probably not, but we can't rule it out yet observationally. The COBE experiment does not have fine enough resolution to see such details. If I smeared this map out in a way that mimicked the COBE resolution, it would be difficult to tell if this was not in fact the way our sky actually looked, at least by eye. Scientists don't trust their eyes and instead perform cautious statistical comparisons with well-defined likelihoods and measures. We do that now, but there is so much uncertainty in our assumptions that we can never really be sure. In the next few years the satellites MAP and *Planck Surveyor* are being launched and they will have much higher resolution. With any luck we'll be able to see better if such patterns persist.

**17 JULY 2000**

Being a physicist encourages defensiveness. We're taught to hang caveats and qualifiers on every statement to protect the proverbial arse and sometimes the literal arse. We're secretly writing for our colleagues, even in our popular articles, or at best are agonizingly aware of their probing eyes. I know I'm supposed to say something about how there are many significant contributions to this field by many talented people which I have left out because of the personal nature of this diary, or plain ignorance. The emphasis on my own work is inevitable and not intended as a misrepresentation. But that's obvious, isn't it? There are people I don't have stories about but who have influenced me, like George Ellis, Helio Fagundes, J. Richard Gott III, K. Taro Inoue, M. Lachieze-Rey, Jean-Pierre Luminet, Angelica de Oliviera-Costa, Boud Roukema, Alexy Starobinsky, Reza Tavakol and Jean-Phillipe Uzan, and others I may have temporarily forgotten. Naturally, my impressions of those I'm close to are stronger and I can't help but recount their influence. And so I persist unapologetically, or was that an apology? Here I'll mention Joe Silk and how he initiated some of the first attempts to measure topology. I worked with Joe for years in Berkeley and know him well. Joe with Daniel Stevens and Douglas Scott realized that for the simplest compact space, the flat hypertorus, there would be no large-scale fluctuations in a universe that was too small. There would be no very low notes in a sense. Oddly enough, our sky as seen by COBE does not have very low notes. This may be a random coincidence or it may not. No one is going to say that the missing low harmonics are proof of a compact topology (Stevens, Scott and Silk argued that the data from the COBE satellite showed no conspicuous evidence in favour of a very, very small universe), but I'll admit a few curiosities were piqued. If we try to match this dip in the observed hot and cold spots, it seems likely that if the universe is finite and small, it is probably comparable to the size of the observable universe. This is another one of those uncomfortable coincidences. Why, if the universe is small, should it just happen to be precisely small enough for us to see? It's the same question we ask about curvature: why, if the universe is curved, should it just happen to be precisely curved enough for us to see? I don't know. Maybe even if it is curved, the curves are too big to see. Maybe even if it is compact, it's still too big to see. This wouldn't surprise me. But we can't stop looking until we know.

Various bounds were put on the size of the universe with the COBE observations, and more recently J. Richard Bond, Dmitry Pogosyan and Tarun Souradeep used another method to argue that if the universe is flat and finite it must be just bigger than the observable horizon. In other words, if the universe is flat and has a limit, we won't ever see it.

So, flat finite universes have fallen out of favour in the past few years. Compact negatively curved universes, however, are all the rage. First of all they're difficult to manage and, having the peculiar personality traits associated with most theorists, we like difficult.[2]

There is a major problem trying to understand the musical score in a compact hyperbolic universe. In fact it's impossible to find a simple mathematical expression for the natural harmonics on a compact, negatively curved space. We can imagine looking for the natural modes by wobbling the space, like hitting the drum, and letting the frequencies play themselves out.

With a simple string instrument, the string strung straight and then anchored at two endpoints, we can predict the waves and thereby the sounds they'll emit. Imagine a string instrument so complex, with strings strung on curves and with such intricate conditions at the ends of the string that we could never predict the natural harmonics that instrument supports. We wouldn't be able to solve the complex and unexpected notes. Even if we tried to play two notes nearly identically, any small differences would quickly become conspicuous. We could never play precisely enough, with enough mastery, that we could play the same song twice.

This lack of predictability in the outcome of our strumming is a classic sign of chaos. And in fact, compact hyperbolic spaces support chaotic flows of waves and light and matter. Actually, this is why I became interested in these spaces in the first place. We cannot write down an understanding of such a chaotic instrument, but we could just pluck the thing and record how it played. This is how some people have tried to tackle an understanding of the harmonics on a compact negatively curved space. They numerically simulate the geometry on the computer and shake the space to see the natural modes ring out. Those modes are then recorded as a list of numbers and images. Do we get cacophony, white noise, noise of any kind? At first glance the numerically simulated spots do look random, like cacophony, if we don't know how to listen, or rather, how to look. But there will be patterns hidden there, if we know how to sift through the noise.

---

[2] If space isn't flat but is instead negatively curved, then there is a mildly comforting connection between curvature and topology. If we are just seeing curvature on the edge of the horizon, the rigidity theorem assures us that we have a catalogue of possible topologies which would definitely be small enough to be visible to us. Whether or not we live in one of those, well who's to say.

**31 JULY 2000**

I locked myself out of the loft again. Again. Needless to say I've done it before. After the last time and a night on the couch in the neighbouring knicker factory, I had enough sense to distribute keys around the work-shops so that someone somewhere would be bound to let me in.

My first expedition out of the house and I barely made it off the fourth floor. Wasn't meant to be. More fatalism and determinism invading my peace of mind. At least I've added meaning to the equations.

I know my life is absurdly chaotic. Don't think I'm trying to impose order on nature as an act of rebellion, retaliation. I accept chaos. I like chaos.

Computers help us where chaos hinders, but there's something miser-ably unsatisfying about simulating thousands of possibilities and com-paring them to the data. Especially, when we know that no matter how many thousands of simulated skies we generate, the number will still be staggeringly small compared to the infinite number of topological possi-bilities. Hence the popularity of some pattern-based searches. The philos-ophy of the pattern-based search is that, instead of trying to predict the sky ahead of time, we will use the future satellite maps to look and see what birthmarks are there. This night sky looks random but maybe it's not. Maybe there are patterned marks lurking beneath the noise. More importantly, maybe some patterns are generic to any finite space.

One of the most ingenious suggestions for a pattern-based search is the identification of circles in the sky proposed by Cornish, Spergel and Starkman. The light striking the earth has travelled billions of light years to get here. Since light travels at the same speed in all directions, each beam of light we intercept has travelled the same distance in all directions since the time light last scattered. This distance effectively defines the extent of the observable universe, our cosmological horizon. We could draw an imaginary giant sphere at a horizon distance from which all of the light we intercept has originated. This hypothetical sphere is called the surface of last scatter (Figure 15.2).

The sphere is not a physical thing, it is just the region of the universe from which light manages to reach us. If we were located a billion light years away, on another planet in another galaxy, we'd be intercepting a different sample of the cosmic background radiation drawn from a slightly different hypothetical sphere centred on our new location. When satellites scan the sky in the microwave they return to us an all sky map of our sphere of last scatter. The maps show the distribution of slightly hot and slightly cold variations in the primordial radiation – they show the distribution of the universe's spots.

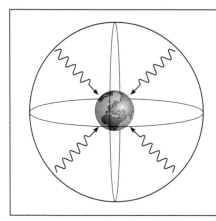

*We receive light from all directions which has travelled 10–15 billion light years.*

*Since light travels the same distance in all directions, this defines a sphere, the surface of last scatter.*

*This cosmic background radiation has carried fossil records from the farthest reaches of the cosmos.*

Figure 15.2 *Primordial radiation from the big bang.*

Imagine the surface of last scatter in a tiling picture of a compact space (Figure 15.3). The surface envelops many copies of a finite universe if the universe is small enough, but none if the extent of space is bigger than the diameter of the surface. We observe the surface of last scatter in one tile and we observe a copy of that surface in a copy of our tile – in fact, an infinite number of copies in an infinite number of tiles. If the universe is small, then some of these copies of the sphere of last scatter will intersect each other. The intersection of a sphere with itself occurs on a circle. If we look in one direction we will see this circle of intersection, but our copy sees this circle of intersection in another direction (Figure 15.4). Since we and our copy are one and the same, we see this circle of intersection in two different directions. This means that the variation in the temperature of the radiation along the two paired

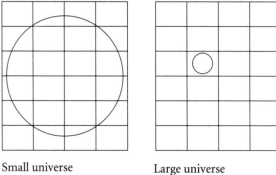

Small universe          Large universe

Figure 15.3 *The surface of last scatter in a tiling picture of a compact space.*

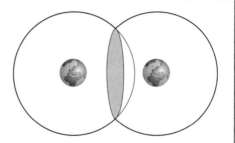

*Two identical copies of the earth receive light from two identical surfaces of last scatter.*

*In the tiling picture, the two spheres can be seen to intersect.*

*The intersection of two spheres occurs along a circle.*

*One copy of the earth sees the circle to the right; the other copy sees the circle to the left.*

*Therefore an observer on the earth must see a pair of identical circles, one to the left and one to the right.*

Figure 15.4 *Circles in the sky.*

circles must be the same. This does not mean the temperature is constant along a given circle, but rather that it varies in precisely the same way along the pair. The cosmic background filling our sky will secretly be littered with these circle pairs.

To be clear, this does not mean that when we look up into the sky we will see pairs of geometric circles. Only when we perform cautious and laborious statistical correlations can we draw out these patterns. Otherwise they remain hidden to the eye. But if we can find circles with a statistical analysis of the data from either the MAP or *Planck Surveyor* satellites, a fair fraction of us will faint. The distribution of circle pairs will allow us to reconstruct the shape of space. If there are no circles, then we know the universe is larger than the 30-odd billion light years spanned by the cosmic background radiation.

## 3 AUGUST 2000

Everyone wants to know what happened between Warren and me. Nothing happened. Everything happened. He doesn't know. I don't know. My work killed us, but I don't know how to quit. It was lunacy and weather. It was power, and powerlessness, inequity and history.

He knows a me that isn't any more. He's my only record of that time.

His memory has the only pictures. My history is a carpet rolling up behind me. I have to run faster to stay ahead of it as it curls up and nicks my heels.

I'm leaning over the top of my new gold 1950s cocktail bar, stocked with multicoloured wine glasses. Since my neighbours keep coming over with multiple bottles of wine, I thought it was about time we started to use glasses. I haven't actually sunk so low as to swig from the bottle, but I can't vouch for some of my neighbores. Oh, a Freudian. Neighbores. I like that. Actually, none of them are bores, they are quite hilarious and amuse themselves and me no end. Nor do any of them swig from the bottle, but we get by with makeshift glasses from old jars. Industrial living bonds us.

I survey the progress of my warehouse conversion. That's it. I'm not going anywhere, I lie. A month later I'm on a plane to California.


**7 AUGUST 2000**

A neighbour wanders towards me along the open catwalk of our industrial site. I'm feeling like I should stew for a bit in a self-imposed silence, but she asks me on my way out if I'm the cosmologist. That's how I'm known in the neighbourhood, as the local astrophysicist. From the catwalk neighbours I don't yet know wave their hellos. She starts saying, so you think the universe is like a polyhedron? I'm not shocked, but what a peculiar and obscure bit of knowledge that is. Where'd she get that bit of information? Most of my colleagues would not connect a finite universe with polyhedra. So we get talking about the interconnectedness of it all, the golden mean, the Fibonacci sequences and biology, and I start to remember the chapter in a book on biology by J. D. Murray that John Barrow once showed me called *How the Leopard Got Its Spots*, a reference to Rudyard Kipling's story of the same title. In 1907 Kipling (1865–1936) became the first Englishman to receive a Nobel prize for literature. A nostalgic imperialist but great storyteller, Kipling wrote classics like *The Jungle Book* and *The Man Who Would Be King*. I glanced at that chapter in Murray's biology text when John took the book from his vast library, which fills his floor-to-ceiling shelves. And, like the hundreds of topics John muses over, it takes up our conversation. We throw it around a bit and set it down again. I was visiting him in Brighton, a couple of years before I moved there. I stayed in this bed and breakfast in the rain that all astronomers stay in, both the rain and the B & B. It was a sweet, melancholy couple of weeks. I'd look out of my

Figure 15.5 *Two points may appear far apart but on a finite space might actually be close together.*

window and see the Ferris wheel lit up on the Brighton pier and a damp carpet flowed from the ocean to my windowsill.

That little seed of mathematical biology was forgotten for a while. But that glance at the page in a book and the one out of our hundreds of conversations simmered. When it was time to get back to new ideas on topology, it was just there. An obvious parallel. The universe imitating life. I wrote to John some many months later and we wrote a paper with Joe Silk and some talented students, Evan Scannapieco and Giancarlo de Gasperis. My boys, I like to call them. Evan came up with some lovely ways to visualize the patterns I had been groping for and we wrote a technical article called 'How the Universe Got Its Spots'.

We can think of a finite space as a complex lens or a peculiar mirror, redirecting the light. If we see a hot spot in the sky and another in the opposite direction, there is no reason to expect them to be related. They come from a distance that today is separated by nearly 30 billion light years. However, if the universe is finite light from opposite directions might in fact originate from the same point (Figure 15.5). So not only are they related, they should be completely identical.

We can exploit the ghost images in the sky to filter out patterned maps of the sky. We can place a bright spot on our map if the temperature of the cosmic background radiation in that direction is identical to the temperature in exactly the opposite direction, in other words if the point and its opposite are ghost images of each other. We call these maps antipodal and here's one of my favourites in Figure 15.6. The figure reflects one space from a group of extremely regular spaces constructed by the systematic identification of the faces of a regular icosahedron. These are known as the Best spaces not just as a superlative but also because the mathematician who discovered them was coincidentally named Best. In the 1970s L. A. Best found three distinct topologies by finding three distinct sets of rules for identifying the faces of the icosahedron. The antipodal map of Figure 15.6 is an idealized version of what we could expect if the universe was shaped like a Best space and we could make perfect observations with the future satellites. The antipodal

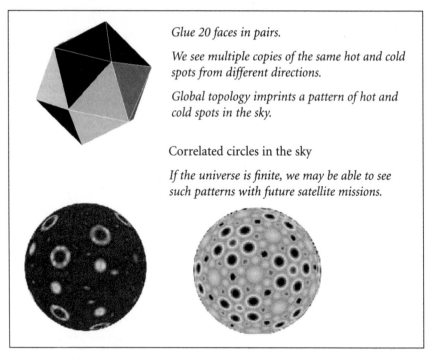

*Glue 20 faces in pairs.*

*We see multiple copies of the same hot and cold spots from different directions.*

*Global topology imprints a pattern of hot and cold spots in the sky.*

Correlated circles in the sky

*If the universe is finite, we may be able to see such patterns with future satellite missions.*

Figure 15.6 *The Best space is a compact icosahedron.*

map finds any ghosts in the sky which are diametrically opposite and has located circle pairs.

There are other correlations we can look for too. We can compare one hot spot to the rest of the sky, and locate all the ghost images of that one spot. Actually Evan found this by accident with a bug in his code. The mistake gave us a good picture but in answer to a different question from the one we intended to ask. We realized it wasn't a bad way to search for repeated images of any given point. The pattern of ghost images traces out aspects of the geometry. There are hexagons and triangles and several orders of symmetry nested in the map (Figure 15.6). Satellite observations of the cosmic microwave sky will certainly look random at first pass. What these maps tell us is that there are patterns buried there, patterns that can be drawn out by locating the ghost images in the sea of hot and cold spots.

We can predict the patterns of these maps for any known topology. The smallest known negatively curved space is the Weeks space. An antipodal map in this geometry would look to us as shown in Figure 15.7. Each space produces its own signature patterns, with the Thurston space producing a series of unusual marks in Figure 15.8. Even though

Figure 15.7 *Left: The fundamental shape of the Weeks space. The space is rendered compact by identifying the faces of the polyhedron. Equivalently, one can imagine tiling a negatively curved three-dimensional volume with identical copies of this shape. Right: A computer simulation of a pattern that would be hidden in the cosmic background radiation if we lived in the Weeks space. The pattern is drawn out by comparing opposite points around the sky and marking the sphere with a bright spot if the opposite points have the same temperature and marking the sphere with a dark spot if opposite points have different temperatures.*

Figure 15.8 *Left: The fundamental shape of the Thurston space. The topology is finite and can be constructed by identifying like faces of the polyhedron. Equivalently, a negatively curved three-dimensional space can be tiled by identical copies of the fundamental polyhedron. Right: The unique pattern that would be encoded in the cosmic microwave background if we lived in the Thurston space. Light from different directions in the sky would be clone images of other directions in the sky as a result of the multiconnected topology. The bright marks identify points in the sky which would be repeated images of the sky in the opposite direction.*

flat spaces, like the compact hexagonal prism, aren't the most popular, we can still muse over their unique patterns. Hexagons of different sizes would lead to the different patterns of Figures 15.9 and 15.10. We could fill a cosmic zoo with the varied markings of the universe's possible spots. Our universe would be there in the zoo somewhere. Using data from future satellite observations, we can search for ghosts in the maps

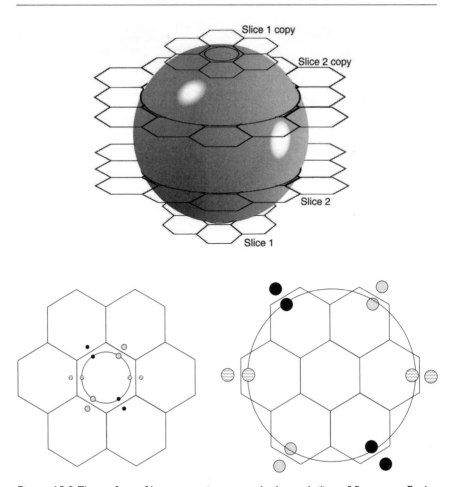

Figure 15.9 *The surface of last scatter intersects the layered tiling of flat space. Each full tile represents a copy of the fundamental domain. Slice 1 is represented on the lower left and slice 2 on the lower right. The dots show correlated points picked up in the antipodal map and explain the emergence of the hexagonal geometry in Figure 15.10.*

and then reconstruct the geometry and topology of the universe. We can identify our universe in the zoo of possibilities.

The universe's radiation spots are reminiscent of the leopard's spots. The processes forming the cosmic patterns share similarities with those forming the patterns on the coats of mammals. In the late 1970s and early 1980s, the mathematical biologist J. D. Murray studied how the leopard might have got its spots through a kind of pattern formation. Murray built on the ideas of the great English mathematician Alan

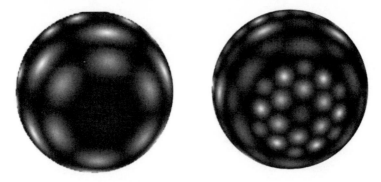

*Figure 15.10 These are orthographic projections of the antipodal maps for a hypothetical universe in the shape of a hexagonal prism. The difference between the map on the left and the map on the right is the size of the universe. In the figure on the left, the space is fairly large compared to the observable universe. In the figure on the right, the space is considerably smaller, so a large number of clone images become visible.*

Turing (1912–1954). History tells us that Turing was tried and convicted of homosexual activity. He came forward to the police to pre-empt any accusations against him. The state subjected him to injections of oestrogen as an alternative to a prison sentence – quite outrageous. Dosed up with elevated oestrogen levels, he drove his mathematical interests forward. Turing is famous for his contributions to code breaking and he continued to play an important part in government studies until he was forcibly withdrawn from intelligence activities. As far as I know, the reason he lost his security clearance remains a secret. Turing died of cyanide poisoning after he found himself under some level of government surveillance. Suicide? Accident? Turing bit from the forbidden fruit, specifically an apple laced with cyanide, and took his wisdom to his premature grave.

Turing's ideas on the flow of chemicals were utilized by Murray to understand the zebra's stripes. Certain elements of this mathematical biology apply to cosmology. There are certainly differences between a mammal and the cosmos. There are processes critical to the formation of animal patterns which are not at work in the cosmos. Still, there are striking geometric similarities. As mammals develop in embryo, they're bathed in chemicals that determine the future markings of the animal's coat. The chemicals fluctuate between high concentrations and low concentrations, effectively establishing a fluctuation wavelength. Regions of high concentration predispose cells to produce a dark stain on the

*Figure 15.11 How the leopard got its spots, and the badger.*

animal's coat, while low concentrations leave the cells and the subsequent fur unmarked. The geometry and size of the developing embryo set the pattern in the chemicals responsible for pigmentation, just like the geometry and shape of space can set the pattern of hot and cold spots in the cosmic background radiation. If the animal is small compared to the distance over which the chemicals vary, the animal will be uniform in colour. A bigger embryo creates two-toned critters like badgers and goats, and bigger still the beast is born with a dark face, white torso, then dark behind, like an anteater. The bigger the animal gets, the more detail in the coat, with spotted giraffes as some of the largest mammals that still flaunt a coat pattern. Bigger beasts, like elephants and hippopotami, go over to being uniform in colour again. Geometry is just as important as scale. The leopard's spots can be imposed by a cylindrical embryo, while the zebra's stripes are drawn out by the tapering of a conical geometry (Figure 15.11).

Mathematical biologists already suspect that the chemical fluctuations which imprint birthmarks on the animal kingdom may have mathematical similarities to the waves on a drum. And mathematical cosmologists already suspect that the waves on a drum may have mathematical similarities to the waves in the curvature of spacetime, which imprint birthmarks on the cosmos. Linking cosmology directly to biology, a zoo of universes with different geometries and topologies could replicate all of the animal markings from zebra stripes to leopard spots.

## 19 SEPTEMBER 2000

I was just at a public conversation between Lee Smolin and the artist Marc Quinn at the Institute for Contemporary Arts in London. My neighbour Ben calls Quinn 'blood-head' in homage to the head that Quinn cast in his own blood. So blood-head and Lee were discussing how a conscious entity can be made up of living cells, like blood cells, which themselves are not conscious. Our blood cells, skin cells, heart tissue, are not individually self-aware even if they are alive by certain criteria. Somehow these living cells collectively form a sentient being – a person. Lee suggested that maybe the entire planet with its individual living cultures and ecologies is one giant organism and we'd be no more aware of its consciousness than blood cells are of ours. Too frightening to bear thinking about.

Imagine millions of conscious little blood cells trying belligerently to understand the body they float through. Here we are, little and futile, trying to understand things that very well could be beyond us. But we are tenacious. Satellites are pending and some of us will stop and dig through the data to see if we can find patterns and circles in the sky.

We are not guaranteed success with any of our approaches. Statistics have their shortcomings. Patterns may get distorted with imperfections in how the universe functions as a geometric lens. We might just miss the images in the grainy imperfections of the data. Or the universe may just be really really big, too big to ever see the topology. But the beauty is in the trying and hoping and striving, even if that's poor consolation for sacrificing our personal welfare.

Still, despite our incredible physical limitations, from this restricted predicament, bound to our puny solar system, we may well see the entire shape of space. From the earth's shores, the light from the big bang washes over us encrypted with geometric patterns, like coded messages in a giant bottle.

# 16

## THE ULTIMATE PREDICTION

**29 SEPTEMBER 2000**

I'm in California, beautiful glorious California. I live in a collision of a slum and paradise. Despite the beauty of our backyards, our front yards are hedged in by acts of unexpected violence. Someone got mugged last night, hit with a pipe and the way he tells the story we can't help but laugh from nerves and an acceptance of the war zone we live in and nonetheless love. The victim asks the mugger for a receipt so he can deduct the loss from his taxes.

Every night it seems we collect in the garden and someone else has cooked an abundance of gorgeous food. Every ingredient is carefully selected and succulent. There's Persian food and California cooking that would make you weep from happiness. The garden is beautiful and overrun and everyday someone else tends the flowers.

I needed some time to myself, just to charge up, be quiet so writing would be a release and not a chore. Last night I climbed into pyjamas at 6.30 and planned on submerging myself in that heavy silence. But then in they came, one after the other, wet from the first rain of the season and cheery with wine, carting bottles of red and food to be cooked and so another party was launched. We laughed and ate like gluttons. So today I've had a late start and am trying to clear the fog from my head. Trying to remember where we left off.

I think we left off here: the universe may be finite but simply too huge to ever let us in on her secret. But we don't have to abandon all hope. The universe keeps wrapping in on itself, at least metaphorically, if not literally. The universe's largest landscape inevitably acquiesces to the laws governing the smallest constituents. The big bang, as the ultimate act of creation, calls on all of those laws. Maybe we should turn our sights inward towards physics on the smallest scales and see if there isn't a reflection of physics on the largest scales.

There is a campaign for simplicity in fundamental science. The simpler, the better. The ultimate achievement for this campaign would be to fuse all of the forces into one. TOEs (Theories of Everything) aspire to explain the entire cosmos with one line of writing, one mathematical sentence. Life feels far more complicated to us. We experience electricity and gravity and nuclear forces, all of which seem distinct. But ultimately all these forces may be different phases of the same thing, just as water, ice and steam are different phases of the molecule $H_2O$. The belief is that all of the fundamental forces are just different manifestations, different phases, of the same force. Theoretical physics has had a great deal of success with this notion of unification, but gravity has been evasive, unwilling as yet to unify with the other forces. Until string theory.

String theory is currently the only theory that can work gravity into a unified framework. The idea began with the suggestion that our smallest constituents are not particles but indivisible strings. As a theory beyond Einstein's, it might be wondered what string theory has to say about the finiteness of space. The model is not well enough understood to make simple predictions for our universe, but predictions are trickling out and wild possibilities have been advanced. String theorists suggest we may not live in three dimensions but rather ten or eleven or twenty-seven. We are oblivious to the extra dimensions because in a sense we are too big to notice that they are there. These extra dimensions are small and, importantly, finite. What about the three we know and love? Are they finite? We cannot make that prediction yet, but we might be able to look inward at the small dimensions if not outward at the cosmic birthmarks.

Theodor Franz Eduard Kaluza (1885–1945) was the first to put forth the suggestion that three dimensions might not be all there is, that things might not be as they seem. He brought his ideas to Einstein's attention in 1919. Kaluza's invention had the first hints of a unification of gravity and particle physics, although it would not ultimately survive the battle against experiment in its most naive form. Kaluza's bold suggestion was given form by the Swedish physicist Oskar Klein (1894–1977) and has been resuscitated again and again in more modern guise. The existence of hidden, curled-up dimensions may be a generic prediction of any theory beyond Einstein's, not just of string theory.

Like the oblivious Flatlanders, we may be unaware of these other dimensions which are everywhere, unable to stick our hands into these extra dimensions simply because they are too small to accommodate our thick appendages. They are small in the sense of being topologically compact, like the three-dimensional straw that has two compact small dimensions and one big expanded dimension.

These small dimensions need not be static. In fact, theorists have largely floundered in attempts to figure out how to keep them from expanding and contracting. Even if we can't see them, or poke our toe in them, we are not immune to the influence of a dynamical internal dimension. A moving internal dimension could alter the expansion of the universe, just as squeezing one direction of a balloon inflates the free direction. Or the shape and size of extra dimensions could change how we perceive the strength of gravity. By looking at these folded-up little dimensions either through the brazen trickery of theory or the even more devout methods of experiment, we might deduce the shape of the large dimensions.

**1 OCTOBER 2000**

The view from the San Francisco hills kills me. The bay and the lumps of earth duck behind the buildings so as not to blind you with magnificence. Avert your eyes, avert your eyes, I feel like saying. I think of the view of London from the top of my building in the East End. It's impossible that I should find these both beautiful. The inadvertent beauty of my industrial British home. I look over London and try to place myself in relation to a map of city lights. I am here. Only now I am not there, I am here, in California.

Nothing is as it seems. Our bodies are mostly water. Water is mostly empty space. Empty space is a harmonic played on a fundamental string. Quantum mechanics says that nature is fundamentally grainy when we focus closely enough. The fundamental grains are made from a handful of different kinds of particles: quarks, leptons, photons, gravitons (which are the quantum units comprising a gravitational wave) and the like. A fairly thrilling theory threatens to overthrow this atomistic assumption. If we looked at the fundamental grains, we would not find point particles, but instead a collective of identical strings. The notes of the string correspond to the different particles that appear to make up the world. So ultimately there is not a handful of distinct particles but only one kind of something, a string. The apparently distinct fundamental quarks and lepton, gluons and gravitons are the varied resonances of these identical strings. Spacetime and matter unify as the intrinsic notes of a complex melody whose score is string theory, the ultimate Theory of Everything.

The description is deceptively simple and seductive. The practice, however, is not. The most direct calculations in string theory are fraught with harrowing pitfalls. String theory succeeds at the nearly impossible – it appears to unify gravity with the other forces – but fails at what might seem the simpler tasks.

String theorists flail in the midst of a plethora of approximations and approaches, but none of the core equations has yet been derived. The landscape of the theory has become so confusing that people forget what their acronyms mean, M-theory being the quintessential example. Choices abound from Manifold to Membrane to Multifarious. Probably the M stands for manifold, but I'm not sure and nor apparently is anyone else. Despite its confounding difficulty, string theory is a beautiful idea and matches the criterion of having a simple plain English expression. It is tempting to believe that one day someone will discover the math to foster a simple formulation.

Strings are so minute they are $10^{20}$ times smaller than an atomic nucleus – and by all current experimental tests would look just like point particles. String theory avoids the sickness of singularities that threatens the health of general relativity. If we tried to create an infinite point-like singularity, we simply couldn't. The more energy pumped into the string, the bigger it would become, so that it actually resists the formation of a singularity and protects the righteous and the good from their depravity (although it is still an unresolved issue if string theory can cure every singularity).

Strings are small enough to move around in small compact dimensions. The extra vibrational modes allowed with this greater freedom are required to eliminate anomalous results from the string theory predictions and leave it a healthier, if still unsolved, theory. The geometry of the internal dimensions shapes the vibrational modes of the string, just as it leaves patterns in the universe's spots. The shape of the small dimensions is therefore constrained by the spectrum of fundamental particles that we live with, although it is still not precisely predicted. The first pioneers expected nine extra spatial dimensions, but this number was later extended to ten for technical reasons. These days the number goes up and down. No real intuition exists for the number of dimensions – whether it's a profound choice or merely incidental.

We are on the cusp of old times and new. String theorists don't bat an eye at an extra dimension and have grown to accept the complex topologies allowed, called Calabi-Yau manifolds. Yet at the same time in history, astronomers sputter at the suggestion. These are just psychological dispositions and again nature taunts our convention, our intuition, our fear. You have to be brave to study her brilliance. If string theory triumphs, it will predict the topology of the large and the small, the number of dimensions, and the geometry of the big bang.

Fundamental physicists are so hung up on symmetry that any break in the harmoniously equitable requires explanation. One conspicuous division introduced to tame string theory is the peculiar separation between the three large dimensions and the small extra dimensions. If there are extra dimensions, then why are three large and the others small?

There is no answer to this question yet, but some intriguing suggestions have been put forth. Strings can wind around the small dimensions constricting them, stunting their growth so that they are roped down into Planck scale sizes. They could in principle constrict all the dimensions with this strangulation, but Robert Brandenberger and Cumrun Vafa showed that in three dimensions two strings had a larger likelihood of colliding with other strings. A string winding one way and a string winding the other can merge, interconnect, so the net effect is an unwrapping of the string. As the strings collide they unwind, releasing three out of the many dimensions from bondage and allowing their unrestrained expansion.

Another suggestion is that we live on a brane, a three-dimensional volume nested in some higher-dimensional space. People like branes, the name derived from the word 'membrane' I suppose, because they get to make all kinds of insider physics jokes, like calling them 'p-branes' or saying they have black holes on the brane. We are confined to the brane, but waves in the shape of space can move through the whole higher-dimensional volume. Our understanding of branes in the context of an evolving universe is still nascent, but it's giving many phenomenologists something to think about.

## 11 OCTOBER 2000

I met a woman tonight who was the perfect audience for my book. In fact, she's probably the only audience for this book. She was speaking so quickly that I realized what I sound like when my speech slurs because the words collide. It was all Cantor and Pythagoras and infinity. Insanity, madness, obsession, math, objectivity, truth, science and art.

These friends always impress me. They're sculptors and tailors, not scientists or spies. I've chosen them with the peculiar attentiveness of a shell collector stupidly combing the overwhelming multitude of broken detritus to hold up one shell so beautiful that it finds its way into my pocket, lining my clothes with sand. And then another. Not too many, so that the sheer numbers could never diminish the value of any one.

Everything fits together. Curved spaces, numbers, light, her, me. We're all somehow complex variations of some simpler score. At least that's the premise of unification. Even though string theory does well with unifying particles, it doesn't completely manage to quantize gravity. The way gravity emerges as a curved spacetime from a collection of strings is hardly clear. In *The Elegant Universe*, Brian Greene described the fabric of spacetime as sewn together out of specific vibrations of fundamental strings. Space and time emerge out of the coherent harmonies of a symphony of strings. Nice imagery he creates. We still don't have a firm enough handle on string theory to try to formally predict the early universe cascade that would precipitate out of string theory. But we can make guesses, fairly learned guesses. We can guess that when the universe was small and fierce enough to conjure up the phantoms of unification, namely strings, there was no clear meaning to space and no clear meaning to time. Each loop of string would ring out gravity waves like little spacetime pieces. A coherent sense of 'here' and 'there' emerges only when the quantum gravity era has tempered. There I go, I said 'when' which implies a time, a before and an after. But I do so cautiously. Maybe 'eventually' is a better word, being event based not time based. Eventually, strings strum together cohesively, creating the sense of a gravitational field and a space to move in and a time to watch pass.

There's another contender for a theory of quantum gravity called loop quantum gravity. The proponents assure me it is a real theory of quantum gravity. Loop quantum gravity, sometimes called quantum geometry, quantizes gravity separately without addressing the problem of unification. The popularity of string theory has been fuelled by the surprising ease of bringing gravity into the fold of unification, even if it does not fully quantize spacetime. So one theory, loop quantum gravity, is a real quantum theory, while the other, string theory, is a real TOE (Theory of Everything). Probably, as Lee Smolin suggests in his book *Three Roads to Quantum Gravity*, they are different paths to the peak of the same mountain. Loop quantum gravity understands that space is quantized into individual chunks of areas and volumes when resolved finely enough. But maybe as a network, the loop quantum gravity states will build up a coherent background spacetime. On this background, some element of string theory may turn out to be true where all of the fundamental forces are the notes played on the strings.

One of the burdens of any quantum theory and any TOE is to explain the eerie connection between black holes and thermodynamics, the statistical mechanics of hot bodies. Both string theory and quantum geometry are able to explicitly derive thermodynamic properties of black holes, such as the entropy and temperature at which they evaporate, which is a triumph for both of those theories.

When we do finally have a TOE that is also a quantum theory, all of our tales will have to be retold. We will really be able to ask incisive and answerable questions about our big bang, about the geometry and topology of space and about our profound role as observers.

# 17

## SCARS OF CREATION

I'm back in London and it's storming. The island fooled me on the first day with an unmistakable appearance of the sun. I leaned back in the black cab home and watched the sun follow us through London. But today it's storming. I hear thunder over the front of the building and I can't help but dwell on the slow progress of the clouds. They are the colour of dirty rain water, as though the city stood upside down and let the puddles collect on the roof of the sky. The weather is manipulative like a combative partner who makes me grateful for any break in the hostility. I savour the good moments with abandon. It must be near noon. There I go, trying to tell the time from the progress of the sun.

The warehouse is changing. New people have moved in during the three months I was gone. It's a psychology test in this building. These units are empty shells when we move in and everyone within the first few months has differentiated themselves with their necessities. Some people have washing machines but no toilets. Some have full kitchen suites but no showers. I have a full bathroom suite but no kitchen. You'd think we'd share more than we do. No one seems interested in using my shower, I don't ask where they go to bathe and I have little interest in cooking, so we're all actually pretty satisfied with our individuated versions of incompleteness.

I had planned to go into the marvels of chaos theory. But I think I have run out of time. It's Christmas today and all I can do is write, holing myself away in my unfinished studio. Looking at the ugly red on the concrete floors and wondering when I'll have time to paint them a glossy white or finish the walls around my Milan bathroom suite. Mark has called me no fewer than five times today to ask me to a Christmas breakfast at one place, then a Christmas lunch at another, then

Christmas champagne chocolates, then two Christmas dinners. Maybe I'll go to dinner which will turn into a party and an all-night celebration. Then there are articles to write, more trips to make, more seminars, more airports and cabs. I love chaos but it's notoriously difficult to define. Open any book on chaos and tell me if any of them define that slick word that graces their covers.

I will say this. The word 'chaos' conjures up an image of hysteria and things out of control. In a sense this is right. Chaos theory is used to describe systems with a loss of predictability, systems too complex for us to predict how they are going to behave. Chaos has found application in everything and anything from traffic jams to crashing stock markets. Peculiar though it may seem, theoretical physics has been the last discipline to acknowledge chaos. Theoretical physics is guided by an aesthetic of simplicity and has sought predictability, symmetry and order, above all. But remarkable order is known to precipitate out of chaos. Fractals are the most striking example. The snowflake that repeats the same structure on smaller and smaller scales is a fractal; a complex, geometrical object that repeats a bulk structure on smaller and smaller scales.

Chaos is a natural consequence of life in a finite space. As light and matter orbit the compact space their paths cross and tangle, forming an intricate fractal pattern. When galaxies do finally form as the universe ages, they can condense along these scars, like dirt collecting in scratches on an otherwise smooth surface. My editor Peter called these rifts that catalyse the formation of galaxies and clusters of galaxies 'Scars of Creation'.

Theoretical physics still considers chaos and complexity to belong to the arena of applied physics and not the arena of fundamental physics. In complex systems the emergent properties of the collective exceed the sum of the parts, just as meaning that emerges from a sentence exceeds the sum of the constituent letters. If I handed you a collection of letters: one a, an e, three i's, etc., it would not have nearly the impact of saying: your hair is on fire. The simple emerging from the complex has happened to us in physics before. Our suicidal mentors, Boltzmann and Eherenfest, left us a theory of thermodynamics that brilliantly teaches us to manage the previously unmanageable. If I want to know the temperature in this room, I don't track each individual atom. Instead I ask about the collective behaviour of atoms and from that a notion as simple as temperature emerges. Specifically, temperature is a measure of the average energy of motion of the collection of atoms.

Maybe this is what black holes are trying to tell us. The connection with entropy and thermodynamics is truly profound. Not only do TOEs and quantum gravity have to explain these peculiar features of gravity, maybe they should be imitating them.

The reductionist believes that every event, no matter how complicated the experience, has as its conductor the one ultimate law of nature. String theory is one contender for the TOE. We are playing out the notes and vibrations, the symphony of that inevitable score. Complexity and chaos emerge not as new laws of nature but as merely the remarkable collection of harmonics of fundamental strings.

In many ways I agree that this must be true if there is an ultimate law of physics. But I can't help but wonder if there isn't a much more radical and deeper role for chaos in theoretical physics. Maybe there are no symmetries, no firm laws, no rigorous order. Maybe our experience of order and the laws of physics is the order that precipitates from complexity. Maybe there isn't an ultimate law, one fundamental symmetry, but instead many, a proliferation of possible laws, and the seeming symmetries that guide our perception of the forces emerge from the collusion of a democracy of quantum theories. Neil Cornish once remarked to me over cocktails at a bar on Haight Street in San Francisco that he thought symmetries might just be a manifestation of self-organized criticality. Maybe he also had something like this in mind.

One of the disturbing developments in string theory is that there is more than one string theory. String theory is not unique. Theories abound with different underlying symmetries, different requirements for the dimensions and shape of space. People are struggling to determine which is the right theory, but maybe they're all right and the appearance of one theory of everything is like the emergence of a temperature from a collection of moving atoms. What if there is not one theory of everything but many? From afar, quantized bits of space and matter conspire to create the illusion of a continuous universe. Maybe, from afar, many theories of everything, conspire to create out of their fundamental complexity the illusion of a simple universe guided by simple laws of physics and based on simple symmetries. If so, when we look closely to resolve particles into quanta and closer to resolve space into chunks of volume, what weirdness might we uncover? Maybe this is a hint as to why quantum mechanics seems so contradictory[1] – why waves can be

---

[1] It is interesting that there is a spacing of energy levels (called level repulsion) in certain chaotic systems which could be interpreted as analogous to the discrete energy levels in quantum systems.

particles and probability reigns. There would be no one truth, not even layers of truth but a complex organization of competing truths.

There are times I believe that gravity should not be quantized, that it should not be unified with the other fundamental forces, but that gravity will emerge out of a radical new picture of the laws of nature. I know I said that geometry was the soul of gravity, but even curved space may just be a limit of a deeper theory. My good friend Lee Smolin would say that causal structures are the soul of gravity, processes and relations. And if we were playing this game in person, we might agree that geometry is a structure we can define as the set of relations and causal structures. This is probably what spacetime is in the sense that it is not really a cloth in my hand. So spacetime isn't a thing but rather a collection of events and their relations. Even more disarming, *we* may not even be things, just events.

Lee answers my random barrage of questions about the loss of symmetries, emergent phenomena and quantum gravity. He tells me there are a few eccentrics out there who have dabbled with similar ideas. Maybe if I have time next year I can try to develop this position more, maybe with Lee's help.


### 28 DECEMBER 2000

Today is a tiring day. Today I myself am a scar of creation. We're all the scars of creation; our thoughts, our pyramids and monuments are the scars of creation.

I haven't left the fourth floor in three days. I've been sealed in this warehouse for nearly seventy-two hours. I finally threw up the steel shutters to reveal the glorious panoramic industrial scene. In the few days of my seclusion a building has grown by two storeys across the canal. Surely I exaggerate, but the monstrosity does seem to be climbing into space with alarming confidence. From my angle the new addition to our canal-side properties is occulting the prominence of the gas tanks that hold a special place in the local urban landscape. They are immense, the gas tanks are. They rise and fall fast enough to watch, which we do in lieu of a setting sun with bottles of wine, leaning over the metal balustrade of our catwalk.

I am taken by the panoply of rusted metal spikes and wet worn monuments to civic engineering. If I strain, I can see the pointed crown of red lights atop the much-maligned Millennium Dome, and Canary Wharf is dressed in a holiday scheme of coloured lights. It's not unusual to see fireworks, holiday or no holiday, flaring over the horizon above the Thames or in the direction of the Docklands. It's also not unusual to see the failed detonations of amateur pyrotechnics fizzle aggressively in the car park below my windows.

Winter is here and damn is it cold. I throw on a coat and run down the catwalk to Ben's studio to see if he's in and wants a tea and run back to start the kettle. I burn my commercial heater with impunity, so that I can nearly tolerate the ambient temperature with little more than a cotton shirt. Ben comes in from his naturally refrigerated studio, bundled in shawls and blankets and warming his hands by the luxury of my gas heater. Approach cautiously, I warn, and reveal the harsh blue-red burn on my leg from an absent-minded leaning on said heater.

I tell Ben about the facial Mark gave me last night. Mark lacquered on the apricot and almond deep cleansing (but moisturizing) mask with professional gentleness. As I lay there looking a mummified fright, wrapped in grandma's knitted blankets and with a yellowish goo on my face, he delighted me with tales of Manchester and otherworld adventures. Mark Lythgoe and I met in the buffeting waves of these science/art meetings washing over London. He is a neurophysiologist who has collaborated with film-maker Andrew Kötting on moving and strange films. There is a storm billowing in London around art/science collaborations[2] I can't fairly represent all of the rich dialogues and interactions that have bloomed, but I've been caught up in some of the questions. Mark and I have gone around the art/science same/different arguments. Art and science are not the same but they may be different sides of the same coin. Even if we never come up with a consistent philosophy of science or definition of aesthetics, not many people get confused over whether they are in an art gallery or a science museum.

Mark is amazed at the persistent differences in how artists and scientists dress. There is a stereotype of the badly dressed scientist, but we don't all wear white lab coats any more, although many of us do wear glasses. I once heard a case argued for a link between myopics and mathematics. Despite myself I have to admit that bespectacled physicists can often be found standing in soiled clothes that don't fit at a dusty board, their hands and backs covered in chalk. Chalk on their shirts and the seat of their pants.

Fashion aside, there is a commonality to the two acutely human

---

[2] There is an interesting book, *Strange and Charmed*, edited by Siân Ede, that chronicles some of the collisions between science and art.

endeavours of science and art, scars of our creation. I once wrote an essay on art and science in a confused attempt to get some obscure funding. I didn't get the funding. The essay was inspired by a book I rediscovered on the table by a friend's bed. I loaned it to him years ago and it's as if no time had passed when I saw it still sitting out as though it were his recent reading.

My friend has a scar across his skinny stomach, is beginning to go grey and is covered with freckles like tiny leopard spots. I suppose it's a crime that I know these things now, a transgression against boundaries of friendships that tangle Warren and I together. But I can't hold Warren in my mind any more, rolling him over like a familiar stone. I woke in the morning to the cover of that old book and didn't touch it. But I did spend the quiet hours of that morning looking at the corner of the gold and black cover that was exposed on the lower shelf of his night stand. I thought about that happily forgotten essay and checked in to see if my ideas had ripened into something better.

The book is *Gödel, Escher, Bach: The Eternal Golden Braid* written by Douglas Hofstadter. Gödel the mathematician, Escher the artist, Bach the musician. Kurt Gödel was born in Austria-Hungary in 1906 (1906–1974). Probably Gödel's most influential contributions to mathematics were his incompleteness theorems. His theories of incompleteness delivered fatal blows to the foundation of mathematics. Greats such as Hilbert were dedicated to the formulation of a mathematics built on simple axioms, simple principles that are taken to be true without proof. Devastatingly, Gödel managed to prove that there are propositions which cannot be proven or disproven within the context of axiomatic mathematics. He proved that some propositions are unprovable. This isn't to say that mathematics is rendered useless but just that mathematics must not be entirely self-contained and not entirely comprehensive. Is it a surprise that he had nervous breakdowns? The first breakdown was provoked when the violence of Hitler's pawns infiltrated his maybe somewhat politically unaware life. I don't know all the details but have heard that the murder of an influential logician by a young National Socialist precipitated his decline. But then I've also heard he was lovesick. Despair followed the murder (or the heartache) and a breakdown pounced on the heels of despair. His life story wouldn't end well. Gödel settled at the Institute for Advanced Study in Princeton in 1953 where depression and delusion would find him. A dread that he was being poisoned grasped his mind. He spurned food only to die of starvation sitting in a hospital chair in Princeton after weeks of self-deprivation. At least he wasn't poisoned. Did he fear Turing had been so murdered? For some reason I find his oblique suicide particularly disturbing.

Hofstadter didn't get ensnared in their personal tragedies but I do remember he argued that 'Gödel and Escher and Bach were only shadows cast in different directions by some solid central essence.' That central essence is human creativity and he tried to convince me through his many hundreds of pages that creativity emerges out of tangled hierarchies of rules and thoughts. He calls them strange loops. Simplistically, I recall that strange loops are tangled paths through levels of rules that eventually end where they began. He finds the strange loop in Gödel's mathematical theorem, the structure of Bach's fugues, the tricks in Escher's drawings. The paradoxical strange loop emerges only out of self-referential systems. We're self-referential. Computers are not. So computers are neither sentient nor animate and have no free will. Hofstadter believes that out of the tangled hierarchy emerges a sense of will and, ultimately, human creativity. I'm not sure how to take this, but I'm happy with any ray of hope that I might find a notion of free will I could believe in without lying to myself. Despite the fear that it strikes in my heart, I still live my life with the persistent illusion of not only free will but also responsibility. I still hold myself accountable, and others. But I don't believe it intellectually. Nor do I *not* believe it. I'm agnostic on the issue of human will and freedom.

## 29 DECEMBER 2000

I know it's been said before that there are evolutionary and cosmic influences on the construction of the human mind and perception. Our penchant for colour or our tendency for pattern recognition is not mysterious given the planet we sprung from. These evolved human perceptions in turn influence our analytic thinking. We grapple with abstraction in mathematics and science by utilizing pictorial representations. Maybe it's just humans perceiving order when order is the exception and not the rule.

Are you still reading Jung? I want to call you and see how your studies are going. I think there's a connection to Jung here and not just because I think all things are connected. My beloved friend Patrick McNamara used to tell me about Jungian theory. He studied with Chomsky at MIT. You know when you're too aware of your own gestation to be confident in your judgement but you think the people you know are spectacular and then a decade goes by and you realize they *were* spectacular. I miss

Patrick and Shep Doeleman, my treasured friend from the MIT days who saved me from insanity by celebrating dementia. I miss their rabid antics in their shared kitchen in Cambridge, Massachusetts. Patrick would tell me about Chomsky and his arguments that language is innate to human development. A universal grammar is built into the mind as a seed that germinates in response to the right conditions and stimuli. We're born with the predisposition to speak any language that can be pared down to something like five elementary languages, or maybe even just one, and we selectively forget those we don't learn.

Jung championed the view that the human psyche is born with a structure. We have two eyes and a nose as a result of genetic evolution and no differently did we acquire a structure to the psyche. The mind has more than just a personal element. There is also a collective character common to all human beings. I went to the city farm in the East End of London. All ducks acted like ducks. There was only one turkey and I can only assume he behaved as all other turkeys would, which was pretty stupidly if you ask me, with this grotesque and inexplicably lewd fleshy globule that hung flaccidly in front of his face. All the hogs acted like hogs. Cute they were too, with giant ears covering their eyes from the feeble sun as they languished in the black mud. And little kids acted like little kids. They seem, I must say, to have been born that way.

The structure of the psyche is innate and also there are innate levels of language. Jung believed there is a language of the unconscious which is fundamentally different from that of the conscious. Myth and allegory are the elements in the language of the unconscious and are used in dreams. Modern abstract language evolved phylogenetically from the more symbolic, archaic language. Maybe early humans literally spoke qualitatively differently from us and the symbolic paintings recording their history are to be interpreted accordingly. What if they really spoke like this: 'sun rises over the buffalo', to mean something like tomorrow is another day?

Are other languages of the mind innate? Is there a language of art and another distinct language of science, each bearing a universal grammar that adjusts to accommodate the influences of culture? Maybe the most abstracted is the language of mathematics and the most symbolic is the language of art. Or maybe it's the other way around. If either is true, any theory of art should respect these ancient lyrics. Many people believe we can't even think outside of language. I'm one of these. If the symbolic expression of dreams and myth requires a more primitive language, I wonder if art requires a similar language and composition. While Escher and even Dali have made explicit reference to modern mathematics, there is the opposite extreme, sometimes my favourite, where obvious conscious meaning is removed from art and the vocabulary of a more primal language is at work. To manage the overwhelming visual information in our everyday lives we symbolize. We assign a name and an abstract concept to the things we see. By dismissing these everyday symbols, another level of symbol surfaces and visual representation can expose a barer meaning, like dreams do.

I guess what I'm getting at is that there are no walls built in the human mind making some of us scientist and some of us artist. They are branches of the same tree, rooted in a common human essence. Maybe it's our ability to step between the different disciplines, weaving strange loops all the while, that's at the core of our creativity. The drive for knowledge must be as innately human as any of our more carnal drives. Maybe the compulsion to ask these questions is rooted in the structure of our minds. Maybe the answers are too.

# 18

## THE SHAPE OF THINGS TO COME

**1 JANUARY 2001**

Writing this as the events unfold is different from writing about the events with the clarity of hindsight. The inane and the mundane are given equal importance in the present before select events acquire special importance in the context of a memory that stretches far into the past and far enough into the future. I will write this without the protection of the future and with the full flavour of the shock and the immediacy of the experience, but I cannot promise I won't revise, edit, reconstruct when a future has accumulated.

Pedro's wife has committed suicide. Catarina was an accomplished artist – a vibrant talent. This isn't distant history any more. These are our lives. None of us knows how to assimilate the tragedy. Our partners follow us on these jarring trajectories like the jagged flock of birds and many of us have watched the light go out in our companions' eyes. Their relationships don't survive. It is impossible to absorb what we're going through, what he's going through, and I don't dare try to explain or understand it.

I say to him, I'm sorry. I am so so sorry. It's a hollow saying and I don't understand what good it does. Rory tells me his theory of the mournful apology. He says everyone wants the chance to forgive the world one person at a time.

Pedro, I am sorry.

**29 JANUARY 2001**

I can't remember if I'm coming or going. East or west. To London or to California. I can recognize the scene from the plane. It's California –

there goes Berkeley, Oakland, San Francisco. From up here the city looks like an infestation. Houses all made out of ticky-tacky and they all look just the same.

I've been back and forth to Berkeley for nearly four months and have hardly seen anyone in my old department. I visit with the other postdocs but avoid the faculty. I'm in hiding, I explain to them, and truth is I've been more productive than I imagined. But I've been moody, so I work quietly, almost in secret. Today I come out of hiding to go to lunch with George Smoot, one of the principal experimentalists on the COBE team. Afterwards we meet up with a crew over coffee in Berkeley's North Side. Sitting around the coffee table, I see myself, as always, on the outside of the inside. My nails are still dirty with charcoal, which escaped my scrubbing after drawing last night. I have this Thursday night ritual that begins with me trying to locate Prudence and our inevitably late arrival to pick up Sara Jane and our spirits, which lift swiftly as we cross the Bay Bridge and enter San Francisco on the way to Diane's studio for her weekly drawing group. All week I look forward to Thursday and I love the whole process. We draw life models for four hours and sometimes stay to talk to Diane, whom we all admire. The danger is to panic and slash out meaningless lines. 'Never make a mark you don't mean,' Diane says as she strolls around the room. Usually Ruth comes too and by some accident we're a group of all women. We intimidate our male models just out of haughtiness, but they get into the collective spirit and are usually quickly befriended. The models display a kind of shy confidence in their nudity. Some nights we have the energy for several drinks in the Mission before finally ambling home.

So here I am today over coffee, looking at the charcoal under my fingers and feeling more the outsider than usual. I look at my colleagues, my friends even. The body language at the table leaves me despondent. Arms crossed, faces stern, torsos twisted. No one looks engaged but more like a dutiful audience as the big boys talk. The faculty assume the disposition of busy men and that's our cue that coffee has ended. They get up and immediately the droves follow, clanking mugs and squeaking chairs. The boys compete for a good position next to their faculty member of choice as we awkwardly file back to the office. I walk with George Smoot whose avuncular advice I enjoy.

Sometimes I wonder what will happen to me. I move further and further outside of academia but I feel happier and happier. My work is better, but I know people worry about me and my choices.

**30 JANUARY 2001**

I'm on the steps of my old apartment in San Francisco, the scene of my yard sales before moving to England. A big giant loop in space and time brings me here again. I guess I wanted to end this where I began, a closed loop. I'm sure I could trace a triangle here.

There's one loop I left for last. I've only talked about compact space, but Einstein taught us that space and time were intimately connected, and they have subsequently been merged into one concept, one word: spacetime. Even time can be made compact. If time is compact, every event will repeat precisely as set by the age of that very peculiar world. In the absurd, you could start on the couch, walk to the sink, pour away the last of the coffee, go back to the couch, walk to the sink, pour away the last of the coffee, go back to the couch, walk to the sink.

Such a short age for the universe is actually absurd if we respect the laws of physics, although a much longer age may be manageable. Our very presence here limits the interval over which time can repeat. We cannot biologically reverse in age and so the interval has to exceed my lifetime or any lifetime. Time can start over again only on a scale set by the largest cosmological forces. Only a universe that can naturally return to its own infancy could be consistent with a closed time loop.

A universe that eventually stops expanding and begins to collapse can in a sense grow young again. If space itself were to collapse, we would be crushed back into the dust of the earth from which we came: the atoms that the sun, the earth and we ourselves stole from a dying star would be returned to space and broken into their subatomic constituents, to the stuff of pure energy where gravity and matter and light merge indistinct. As nature rushes to its demise, the entire universe would become smaller than a discarded speck of dust, and smaller still, and perhaps then it could start all over again. The energy of the implosion would be so great that our cosmos explodes in a big bang, a cosmic rebirth in which space swells and our entire history repeats itself. The same galaxies form and the same stars and planets; and on at least one of those planets there is life. You are born. I am born. Even a proponent of free will can see that, at the very least, we would be limited in the choices we could make. We would live out the same lives, make the same choices, make the same mistakes.

In fairness, our story is amended by the cloud of uncertainty from quantum mechanics. In its origin, the universe becomes a quantum system and it is difficult to know which if any features of the universe would persist. A closed time loop with the added ingredient of quantum uncertainty could, at each big bang, create a universe with slight differences, differences that can develop into ones as slight or as huge as the differences between siblings. In a quantum creation of the universe, different galaxies would form in different locations composed of different stars and new planets. You and I would not be here, but the chances are that someone would. And if we really are our parents' children, they will wonder about the cosmos and will try to find a way to understand their place in the universe.

# EPILOGUE

I'm back in California. Yes, again. The letters that turned into a diary are being turned into a book. Those chapters of my story are finished for me, written and recorded as they are. Peter has the manuscript. I'm walking down my old street. Yes, again. My body is facing south, my eyes are turned east. I'm looking at the old apartment but not stopping. It's been over two years since Warren and I lived there. A year, almost to the day, since we've spoken. I don't even know where he is. I won't slow my stride. I'm moving on. When I turn my head to align my spine and pry the tangle of hair from my eyelashes there's a shadow of a person in front of me. I'm mentally navigating some evasive manoeuvre but there's no time to shift course, no time to recognize that sweet smile, those mad eyes. He's stopped, purposefully blocking my path. Staring at me. Wide eyed. The proverbial deer in the headlights. He's shaking. He's shaken. It's Warren.

# INDEX

3-sphere, 83

Abbott, Edwin Abbott, 107
absolute zero, 90
acronyms, 90
action, 64
Allen, Woody, 82, 113
Alpher, Ralph, 90–91
anteater, 176
anthropic principle, 160
antipodal maps, 171, 172, 175
Aristotle, 7
atoms, 58, 77, 80, 186

Bach, J. S., 190–91
badger, 176
Barrow, J., 7, 12–13, 86, 113, 117, 170, 171
Belinsky, Vladimir, 86, 94, 130
Bell, Jocelyn, 69
Berkeley, 12, 43, 165, 195
Best, L. A., 171
Best space, 144, 171–72
Bethe, Hans, 90–91
big bang, 14, 22, 45–6, 62, 64, 78, 80–81, 84–5, 87, 91, 94–5, 99, 130, 155, 159, 168, 184, 196–7

and black holes, 86
chaotic, 63, 86, 94
geometry, 182
light from, 157, 177
black holes, 22, 44–5, 62, 64–5, 68, 72–8, 86, 99, 102
and big bang, 86
and chaos, 63
evaporation, 77
interiors of, 75–6
properties of, 75
and thermodynamics, 184, 186–7
and universes, 160
Bohr, Niels Henrik David, 58, 61, 163
Boltzmann, Ludwig, 1, 77, 186
bombs, 66–7, 71
Bond, J. R., 165
Bond, R., 117
Bragg, M., 55
Brandenberger, Robert, 182
branes, 182
Broglie, Louis Victor Pierre Raymond duc de, 55
Brouwer, Luitzen Egbbertus Jan, 13

Calabi-Yau manifolds, 182
calculus, 20

Cambridge, 37, 69, 72
Canadian Institute for Theoretical
    Astrophysics, 117
Cantor, Georg Ferdinand Ludwig
    Phillip, 7, 8, 11, 12, 13, 36, 87
carbon, 71
cardinality, 8, 10$n$
CERN, 44
chaos, 22, 63, 64, 86, 117, 166, 186,
    187
Chinese, 69
Chomsky, Noam, 191, 192
Circle line, 66, 67
computer simulations, 166
continued fractional expansion, 9$n$
continuum, 13
Copenhagen Interpretation, 61
Copernicus, 16, 18, 21
Cornish, N., 63–4, 86, 117, 162, 167,
    187
cosmic background radiation, 90,
    92–4, 96, 155, 157, 158, 163, 167,
    169, 171
    hot and cold spots in, 93–4, 155–6,
        163, 164, 165, 176; *see also*
        universe, spots of
cosmic expansion, 83
cosmic microwave background
    (CMB), 90
cosmological constant, 82, 94, 159

Dali, Salvador, 140, 193
Dehn's surgery, 145
depression, 1, 11, 24, 190
    *see also* insanity; madness; mental
        illness
determinism, 21, 60–61, 99
Dicke, Robert, 91
dimensions, 104–14, 121, 123, 125–9,
    132–7, 139–43, 146, 179–80, 182

extra, 179, 181, 182
    *see also* fourth dimension
Doeleman, S., 192
Doroshkevich, Andrei, 91
double-slit experiment, 55
Duchamp, Marcel, 140
dynamics, 19

earth, 6, 16, 17, 18, 41, 44, 64, 79, 81
    orbit of, 26
eclipses, 45
Ede, Siân, 189$n$
Eherenfest, Paul, 1, 61, 186
Einstein, Albert, 5, 15, 24, 27–30,
    34–6, 41–4, 62, 74, 76, 78, 85, 87,
    99, 114, 158, 179, 196
    and black holes, 74
    equations, 66, 83, 86, 100, 105
    and the expanding universe, 82, 94
    and gravity, 35, 38–40, 102
    and insanity, 43
    Nobel prize, 5, 54
    principle of equivalence, 39
    and quantum mechanics, 61
    quantum theory of light, 54, 55
    theories of relativity, 5, 18-19, 22,
        27, 29, 35, 36, 42, 48, 51, 54, 134
    theory of curved space, 42, 44–5,
        64, 100
    *see also* cosmological constant;
        light, speed of; relativity; space,
        curved; spacetime, curved
Einstein–Hilbert action, 64
electromagnetic radiation, 52–3
electromagnetism, 22, 25, 52–3, 62
electrons, 54, 56, 57, 58–9, 68, 69, 88,
    89
electroweak theory, 62
elephant, 176
ellipses, 18–19, 45

Ellis, G., 165
emergent properties (phenomena), 186, 188
energy, conservation of, 58
entropy, 184, 187
Escher, M. C., 140, 190–91, 192
Euclid, 46, 48
Eurocentrism, 21

Fagundes, H., 165
Fermilab, 50–51
Ferreira, P. 43–4, 154, 157, 194
Feynman, Richard, 112
Fibonacci sequences, 170
Fields medals, 117, 145
Flatland, 107–9, 112, 114, 132, 134
fourth dimension, 110–13
fractals, 86–7, 186
Freese, K., 51
Friedman, Alexander, 82–3, 93
fundamental forces, 62, 159, 179, 188

galaxies, 2, 3–4, 44, 73–4, 79–80, 94, 149, 154, 155
    formation of, 90, 186, 196–7
    recession of, 82
Galileo Galilei, 7, 18-19, 20, 21, 68
Gamow, George, 90–91
Gasperis, G. de, 12, 170
gender, 88
geometry, 4, 5, 35–6, 42, 47, 101-102, 135, 140, 141, 143, 188
    of big bang, 182
    computer simulations, 166
    distinction from topology, 119, 132n
    Euclidean, 48
    non-Euclidean, 46, 47–8, 143
    Riemannian, 42, 139
    of space, 45, 48, 83, 94, 100, 149, 184
    of spacetime, 44
    of universe, 97, 136, 156, 157, 173
giraffe, 176
gluons, 180
goat, 176
Gödel, Kurt, 190–91
golden mean (ratio), 9–10, 11
Gott III, J. R., 165
gravitation, 19, 20, 22, 35, 62
gravitational collapse, 67, 68, 72, 75
gravitational lensing, 155
gravitational waves (gravity waves), 74–5, 180, 183
gravitons, 180
gravity, 4, 16, 21–2, 34–5, 38–41, 43–4, 48, 62, 64, 72, 76–8, 88, 99–100, 160, 179, 181, 183, 187–8
    loop quantum, 183–4
    quantum, 77, 87, 88, 102, 157, 183, 187, 188
Greene, Brian, 183
Gribbin, J., 55
Gris, Juan, 140
Grossmann, Marcel, 42
Guth, Alan, 94, 95

Halliwell, J., 67
handles, 120
Hawking, Stephen, 77, 86
Hawking–Penrose singularity theorems, 86, 99
Hawking radiation, 77
Heisenberg, Werner Karl, 56, 57, 58, 61
    see also uncertainty principle
helium, 71, 89
Hermann, Robert, 90–91
Hewish, Anthony, 69

Hilbert, David, 42, 64, 190
hippopotamus, 176
Hobbes, Thomas, 11
Hofstadter, D., 190–91
homogeneity, 83, 85, 93, 94
homosexuality, 24, 174
horizon, 73, 75, 99, 160
'How the Universe Got Its Spots', 171
Hubble, Edwin, 79, 82
   expansion, 81
hydrogen, 71, 89
hyperbolic plane, 48

incompleteness theorems, 190
inertia, 19
infinity, 3, 4, 7–15, 36, 52–3, 77, 99,
      103, 145
   countable, 8, 10, 11
   in nature, 10, 13, 76
   uncountable, 11
inflation, 95–6, 158, 159
Inoue, K. Taro, 165
insanity, 24, 43
   see also depression; madness;
      mental illness
interference, 55
Isham, C., 67, 157
isotropy, 83, 85, 93, 94

Jimenez, N., 118
Jung, C. G., 191, 192
Jupiter, 69

Kac, M., 162
Kaluza, Theodor Franz Eduard, 179
Kant, Immanuel., 2
Kelly, R., 118, 119
Kepler, Johannes, 18–21, 45, 68
Khalatnikov, Isaac Markovitch, 63,
      85, 86, 94, 130, 137$n$

Kipling, Rudyard, 170
Klein, Oscar, 179
Klein bottle, 127, 129, 136, 138
Kolb, R., 88
Kötting, A., 189
Kronecker, Leopold, 11, 12, 13

Lachieze-Rey, M., 165
Landau, Lev Davidovich, 67–8
Lemaître, Georges, 83
leopard, 174, 176, 177
leptons, 180
level repulsion, 187$n$
Lifshitz, Evgeny, 85, 86, 94, 130
light, 27, 29–30, 44, 45, 47, 88–9, 91,
      93, 128-29, 135, 186
   bending of, 64–5, 72
   from big bang, 157, 177
   and black holes, 72–3, 75, 77
   bound to space, 122
   from opposite directions, 171
   particulate nature of, 54–5, 89
   speed of, 26, 27, 28, 30, 31, 32, 34,
      35, 39, 66, 94, 97
   wave theory of, 25–6, 27, 52–5
Linde, André, 159
loop quantum gravity, 183–4
loops, 128, 129, 130
Luminet, J.-P., 165
Lythgoe, M., 156, 189

madness, 1, 12, 90
   see also depression; insanity;
      mental illness
Magueijo, J, 67
Malevich, Kasimir Severinovich, 140
Manhattan project, 71
manifolds, 34, 103
Marx, Groucho, 119
mass, 40–41, 64–5, 66, 67, 83, 100

inertial, 34
mathematicians, 1, 11, 12, 38, 42, 115, 116–18, 144–5
Maxwell, James Clerk, 22, 25, 27, 52–3
McNamara, P., 191
mental illness, 2, 8, 11, 13
  *see also* depression; insanity; madness
Mercury, 18, 21, 45
Michelangelo, 43
Michelson and Morley, 26
Milky Way, 3, 74, 84
Minkowski, Hermann, 5, 42
Möbius strip, 127, 128, 136, 138, 144
Mondrian, Piet, 140
*Monty Python*, 113
moon, 17, 44
Moscow, 63, 124, 137, 147, 150
M-theory, 90, 181
Murray, J. D., 170, 174, 175

native Americans, 69
natural selection, 159, 160
neutrinos, 62, 88
neutron stars, 68, 69, 72
neutrons, 59, 69
Newton, Isaac, 15–16, 18-19, 21, 25, 27, 29, 35, 38–9, 42, 66, 72, 88, 102
  alchemy, 24
  homosexuality, 24
  invention of calculus, 20
  laws, 20, 24, 44
  mental health, 24
  *see also* gravity
Newton's constant, 88
Nobel prizes, 5, 54, 68, 69, 91, 117, 170
Novikov, Igor, 91

nuclear fusion, 66–7, 68, 71, 160
numbers:
  irrational, 9–10, 11, 12
  natural, 9–10, 11, 13
  negative, 11
  rational, 10, 11
numerology, 1

observers, 54–5, 56, 57, 112, 184
Olbers, Wilhelm, 152
Oliviera-Costa, A. de, 165
Oppenheimer, J. Robert, 74
Orwell, George, 7, 109, 110
oxygen, 71

pattern-based searches, 167
Patterson, J., 71
Peebles, James, 91
Penrose, Roger, 85–6
Penzias, Arno, 91
periodic table, 59
phenomenology, 28
photoelectric effect, 54
  experiment, 54, 55
photons, 54, 55, 89, 102, 180
Pi ($\pi$), 9, 11
Picasso, Pablo, 140
Planck, Max Karl Ernst Ludwig, 53–4, 55
Planck time, 88
*Planet of the Apes*, 141
planets, 90, 157, 196–7
Pogosyan, D., 117, 165
probability wave, 56
protons, 59, 160
pulsars, 69
Pythagoreans, 1–2, 9, 12, 110

quanta (quantum particles), 14, 52, 54, 102, 187

quantum theory (quantum
mechanics), 21, 51–2, 54–62,
76–8, 100, 102, 180, 184, 187,
196
  geometry, 183–4
  gravity, 77, 78, 87, 88, 102, 157,
  183, 187, 188; *see also* loop
  quantum gravity
  tunnelling, 58
quarks, 62, 88, 180
quasars, 74
Quinn, M., 177

Randall, L., 87–8
relativity, 24, 28, 31, 32, 35, 43, 51, 77,
161
  general, 5, 15, 19, 22, 35–6, 41–2,
  45, 48, 54, 64, 76–7, 81, 85, 88,
  98–9, 102, 112, 115, 127, 135,
  181
  special, 15, 22, 27–8, 30, 35, 38–9,
  41–3, 54
Riemann, Georg Friedrich Bernhard,
42, 46
rigidity theorem, 146, 166
Robertson, Howard, 82–3
Roukema, B., 165
Rutherford, 58

satellites, 90, 93, 116, 156, 160, 167,
172, 173, 177
  COBE, 90, 93, 164, 165, 195
  MAP, 90, 164, 168
  *Planck Surveyor*, 90, 164, 168
Scannapieco, E., 12, 171, 172
Schramm, D., 51
Schrödinger equation, 57
Schwarzschild, Karl, 45, 64, 74
Scott, D., 165
Segal, Prof., 81

Silk, J., 12, 117, 154, 165, 171
singularities, 72, 73, 76, 85, 86, 99,
102, 181
Slade, Henry, 113
Smolin, L., 55, 87, 159, 160, 177, 183,
188
Smoot, G., 195
social constructivism, 56
Sokolov, I., 117
Souradeep, T., 165
space, 4, 25, 27, 30, 36, 40–41, 51, 86,
94, 183, 196
  beginning of, 81
  curvature of, 77, 157
  curved, 4, 5, 22, 35–6, 40–42, 44,
  46, 93, 101, 127, 135, 143, 161,
  188
  dimensions of, 103
  fabric of, 76, 77
  finite, 144, 186
  flat, 47, 83, 84, 97, 132, 166$n$, 172
  geometry of, 45, 48, 83, 94, 100,
  149, 176, 184
  limit to, 102
  negatively curved, 48, 83, 97, 132$n$,
  139, 143, 145, 146, 166$n$, 172
  positively curved, 47, 83, 97, 132$n$,
  143, 145
  ripples in, 75
  shape of, 100, 145, 146, 168, 176,
  177, 187
  smooth, 163
  topology of, 160, 184
spacetime, 33, 36, 44, 46, 48, 72, 82,
99, 157, 180, 183–4, 188
  contracted to a point, 80
  curvature of, 65, 160, 176
  curved, 39–41, 43, 62, 64, 76, 78,
  100, 183
  diagrams, 112

equivalent to 'universe', 46
geometry of, 44
ripples in, 74, 93
Spergel, D., 162, 167
Stalin, 67, 68
Starkman, G., 117, 162, 167
Starobinsky, A., 165
stars, 65, 67–75, 81, 87, 152–3, 154,
157, 160, 196
collapsed, 65
death of, 65, 68, 72, 73
dense, 64
formation, 65, 67, 90, 196–7
see also black holes; neutron stars;
supernovae; white dwarfs
statistical mechanics, 1
Stevens, D., 165
string theory, 179–84, 187
strong force, 62
suicide, 1, 2, 175, 186, 191
Sumerians, 66
sun, 16, 17, 18, 25, 41, 44, 45, 52–3,
64–5, 66
superfluidity, 68
supernovae, 68–9
surface of last scatter, 167, 168, 169,
174

Tavakol, R., 165
Theories of Everything (TOEs), 90,
179, 181, 183–4
thermodynamics, 184, 186–7
Thorne, K., 99
Thurston, William, 145, 146
Thurston space, 146, 173
tiling, 123, 132, 134–45, 168, 174
time, 25, 27, 30, 36, 40–41, 51, 82,
183, 196
beginning of, 81
closed loop, 196

compact, 196
curved, 44
as dimension, 33, 64, 112
non-existence of, 33, 112
warped, 40
time of last scattering, 89, 163
topological lensing, 155
topology, 5, 14, 34, 48, 62, 64, 97,
101–103, 107, 111, 116–17,
120–23, 129, 134, 138, 140,
145–7, 149, 161, 172
distinction from geometry, 119,
132n
early measurements of, 165
of large dimensions, 114
of space, 160, 184
and string theory, 182
of the universe, 4, 63, 102, 115,
119, 130, 136, 146, 158, 173, 177
transfinite arithmetic, 8, 10n, 36
Traweek, S., 24
triangles, 46–7, 48, 139, 172
Turing, Alan, 174–5, 191
twins, 30–32, 111, 126
Tyco Brahe, 17, 18, 21, 68

uncertainty principle, 56, 58, 61, 68,
77, 93, 163
unification, 183
universe, 48, 51, 76, 79, 83, 94, 101,
104, 125, 127, 132, 143, 160, 197
age of, 196
ageing, 4
beginning of, 4, 64, 85, 87–8, 90,
157
collapse of, 196
compact, 145, 163, 166
cooling, 89
curves of, 127, 135
early, 62, 86, 89, 95, 163, 183

equivalent to spacetime, 46
evolution of, 64, 182
expanding, 4, 64, 79–80, 82, 83–5,
    87, 89, 94, 97, 158, 196
expansion of, 180
extent of, 101
finite, 3, 4, 7, 14, 48, 83, 98,
    116–17, 135, 137, 141, 147, 153,
    156, 158, 161, 163, 165–6, 168,
    171
flat or nearly flat, 158–9, 161,
    165–6
geometry of, 97, 136, 156, 157, 173
infinite, 4, 5, 6, 14, 34, 82, 87, 97–8,
    99, 102, 116–17, 135, 161
inflationary, 159
largest attributes of, 158
negatively curved, 158, 161, 166
observable, 159, 165
positively curved, 158, 161
properties of, 100
recollapse of, 97, 158
shape of, 46, 115
smooth, 95–6, 158
spots of, 167, 171–4, 181; *see also*
    cosmic background radiation,
    hot and cold spots in

static, 82, 83
string theory predictions for, 179
topology of, 4, 63, 102, 115, 119,
    130, 136, 146, 158, 173, 177
Uzan, J.-P., 165

Vafa, Cumrun, 182
van Eyck, Jan, 47
Venus, 68
vision, 67, 70

Walker, Arthur, 82–3
wave–particle duality, 52, 54–5,
    187–8
weak force, 62
Weeks space, 146, 172, 173
Weeks, J., 144, 146, 147
Wheeler, John Archibald, 65, 75, 112
white dwarfs, 68, 69
Wilde, Oscar, 91
Wilson, Robert, 91
wormholes, 64

zebra, 175, 176, 177
Zeno, 13–14